The Inner Planets

The Inner Planets

New Light on the Rocky Worlds
of Mercury, Venus, Earth,
the Moon, Mars, and the Asteroids

Clark R. Chapman

Charles Scribner's Sons / NEW YORK

Copyright © 1977 Clark R. Chapman

Library of Congress Cataloging in Publication Data

Chapman, Clark R.
    The inner planets: new light on the rocky worlds of Mercury, Venus,
Earth, the Moon, Mars, and the Asteroids
        Includes index.
        1. Mercury (Planet) 2. Venus (Planet)
3. Earth. 4. Mars (Planet) I. Title.
QB611.C44   559.9   76-58914
ISBN 0-684-14898-6

1 3 5 7 9 11 13 15 17 19 H/C 20 18 16 14 12 10 8 6 4 2

Printed in the United States of America

Dedicated to the memory of my father
DR. SEVILLE CHAPMAN
who built a two-inch refracting telescope
and, with it, introduced me to the planets

Contents

List of Illustrations

SOME IMPORTANT LUNAR AND PLANETARY SPACE MISSIONS

Spacecraft	Nation	Launched	Target	Arrived	Description
Mariner 2	U.S.A.	8/62	Venus	12/62	Measured temperature, magnetic field
Ranger 7	U.S.A.	7/64	Moon	8/64	First close-up photographs, before impact
Mariner 4	U.S.A.	11/64	Mars	7/65	22 close-up photographs
Venera 3	U.S.S.R.	11/65	Venus	3/66	First impact on another planet
Luna 9	U.S.S.R.	1/66	Moon	2/66	First lunar soft landing
Surveyor 1	U.S.A.	5/66	Moon	6/66	First U.S. soft landing on moon
Lunar Orbiter 1	U.S.A.	8/66	Moon	8/66	First of a series of lunar orbiters to study potential Apollo landing sites
Venera 4	U.S.S.R.	6/67	Venus	10/67	Made measurements during descent through atmosphere
Mariner 5	U.S.A.	6/67	Venus	10/67	Remote measurements of atmosphere
Surveyor 5	U.S.A.	9/67	Moon	9/67	Measured chemical composition of lunar surface
Mariner 6	U.S.A.	2/69	Mars	7/69	Martian fly-by
Mariner 7	U.S.A.	3/69	Mars	8/69	Companion spacecraft to Mariner 6
Apollo 11	U.S.A.	7/69	Moon	7/69	First manned landing, brought back lunar rocks
Apollo 12	U.S.A.	11/69	Moon	11/69	Second manned landing
Luna 16	U.S.S.R.	9/70	Moon	9/70	Unmanned landing, brought back lunar rocks
Apollo 14	U.S.A.	1/71	Moon	2/71	Manned landing in Fra Mauro region

Spacecraft	Nation	Launched	Target	Arrived	Description
Mariner 9	U.S.A.	5/71	Mars	11/71	Relayed pictures and data from Mars orbit for about a year
Apollo 15	U.S.A.	7/71	Moon	8/71	Manned mission to Hadley Rille
Luna 20	U.S.S.R.	2/72	Moon	2/72	Unmanned; sample brought back
Pioneer 10	U.S.A.	3/72	Jupiter	12/73	Traversed asteroid belt, studied Jupiter and its environment
Venera 8	U.S.S.R.	3/72	Venus	7/72	Transmitted data from surface of Venus for 50 minutes
Apollo 16	U.S.A.	4/72	Moon	4/72	First manned landing in lunar uplands
Apollo 17	U.S.A.	12/72	Moon	12/72	Final manned lunar mission
Pioneer 11	U.S.A.	4/73	Jupiter Saturn	12/74 9/79	First man-made object on an escape trajectory from the solar system
Mars 4 and 5	U.S.S.R.	7/73	Mars	1/74	Mars orbiters
Mariner 10	U.S.A.	11/73	Venus Mercury	2/74 3/74	First close-up photographs of Venus and Mercury
Venera 9 and 10	U.S.S.R.	6/75	Venus	10/75	First photographs from surface of Venus
Viking 1	U.S.A.	8/75	Mars	6/76	First measurements from surface of Mars
Viking 2	U.S.A.	9/75	Mars	8/76	Companion mission to Viking 1
Luna 24	U.S.S.R.	8/76	Moon	8/76	Returned two-meter-long soil core
Mariner 11 and 12	U.S.A.	8/77	Jupiter Saturn	1979 1981	One spacecraft may continue on to study Uranus
Pioneer 12 and 13	U.S.A.	1978	Venus	1978	Orbiters with atmospheric descent probes

Preface

"IT'S JUST INCREDIBLE to see that Mars is really there!" Seven years to the day after a man first walked on the moon, Dr. Tim Mutch was savoring the most rewarding moment of his career. Of course Mars was there. Moreover, the Viking spacecraft camera was working! Tim Mutch had been responsible for the team of scientists and engineers who had designed, built, and tested the camera that was at this moment sitting on the orange plains of Mars transmitting to Earth the first close-up of a Martian desert. He was sharing his personal triumph with millions of viewers, watching over his shoulder via network television, as the sliver of a picture gradually expanded into a full-size portrait of the ground beneath Viking's footpad.

The second picture began to develop, left to right—an eye-level panorama of a boulder-strewn scene. Viking scientists were thrilled as they watched their TV monitors at the Jet Propulsion Laboratory. Yet something was wrong. The sky was bright; indeed, as later color pictures revealed, it was bright ochre. Where was the blue-black sky that scientists had expected?

This was to be only the first of many surprises for Viking scientists as the spacecraft's sojourn on Mars lengthened from hours to days, then months. Elated when the three biology experiments seemed to indicate abundant life on the Red Planet, researchers were dumbfounded when the chemistry experiment failed to prove the existence of any organic molecules on Mars. The whole issue of Martian life was cast into a state of ambiguity. Meanwhile, overhead the first Viking Orbiter, later joined by a second, determined that Mars's presumed dry-ice polar caps are in fact composed of water-ice. Later the Orbiter cameras photographed mysterious grooves on the largest of the planet's two moonlets.

The first months of the Viking mission epitomized our ever changing ideas about Earth's ruddy neighbor in space. Ancient associations of Mars with a warlike deity, overprinted by Percival Lowell's turn-of-the-century theories of a canal-building Martian civilization, culminated before World War II in Orson Welles's terrifyingly real radio drama about an invasion from Mars, which few listeners then doubted was teeming with life.

On July 30, 1965, there was again news from Mars. The 18-point headline on the front page of the *New York Times* reported scientists' pessimistic conclusions from the first Space Age reconnaissance of the planet:

MARINER 4'S FINAL PHOTOS
DEPICT A MOONLIKE MARS

Signals radioed across millions of miles from the 575-pound spacecraft had been reconstructed by NASA computers into fifteen close-up pictures of a crater-scarred landscape.

Public fascination with Mars died a death as bleak as the "lifeless" landscape shown on the wirephotos, and the nation's concern was directed inward to a divisive war and social upheaval. A few years later a vice-president suggested sending men to Mars by the

1990s; a bubble of public indignation arose over the outrageous cost of such folly and subsided again.

Meanwhile, oblivious to the death sentence they had imposed, scientists struggled to unravel the secrets of the Red Planet. Bigger and better spacecraft were launched toward the "moonlike" world. The fifteen close-ups became dozens, then thousands. The surface of Mars was found to be covered not only with craters, but also with volcanoes, canyons, and hundreds of river valleys, testifying to a more clement climate in Mars's past. Work proceeded on developing an unmanned mission to land on the Red Planet in order to learn about Martian biology.

In America's bicentennial year news came once again from Mars. The Viking lander stuck out its mechanical arm and gobbled up the dusty orange soil of the Chryse plains in the hopes that Martian microorganisms would chemically reveal their presence. After studying the data telemetered back to Earth, scientists once again issued their preliminary findings to inquisitive reporters. But the data ultimately were ambiguous because Mars is not the precise laboratory environment imagined by the designers of Viking's experiments. Nor would the erstwhile Martian organisms fit precisely the models of recognizable life forms that biologists had constructed from their Earth-biased perspective.

All of which is not to say that the question of life on Mars will never be answered. It will—if not this year, maybe next year, or maybe in a decade or two. Nor does the fact that the question has often been answered prematurely in the past detract from the significance of learning more about our biological heritage and the uniqueness of our ecological niche in space called the planet Earth. Indeed, that the cultural tool we use to seek answers to such questions—the sociological process called science—is a thoroughly human activity of idiosyncratic individuals, each with his or her own insights and shortcomings, should be reassuring in this age when machines and computers threaten us.

If we had known what to expect from Viking, there would have been no point in going to Mars. We should not be confident in future conjectural interpretations or "models" for Mars eventually to be settled upon by Viking scientists; we must return to Mars again and again. As a baby tentatively and imperfectly learns about its environment, so science slowly evolves an ever more accurate understanding of the planets and of our planet's place in the universe. In this book I will guide the reader toward understanding the Earth and the other Earth-like planets in the inner solar system, always from the scientist's perspective, where hypothesis dominates fact.

For assisting me in preparing this book, I wish first to thank every man, woman, and child in America, each of whom has contributed in taxes about a dollar a year to sending unmanned spacecraft toward the planets and about a dime a year to me and all my scientific colleagues so we may understand the data the spacecraft send back. Nearly all of the lovely pictures that illustrate this book are the results of the same dollar per year per person.

I thank Joseph Ashbrook for suggesting I write this book. Lonny Baker, Jennalyn Chapman, David Morrison, and Laurel Wilkening provided especially helpful critical reviews of the manuscript. Early suggestions by Kenneth Heuer and later detailed criticisms by Doe Coover at Charles Scribner's Sons have proven invaluable. Innumerable colleagues have helped to shape the view of the planets reported in this book. Several have contributed illustrations, including William Baum, Michael Belton, Richard Goldstein, William Hartmann, Stephen Larson, and Robert Strom.

1

Planetologists

WHY ARE WE FASCINATED by the planets? Perhaps it is because they are remote, exotic worlds, with vistas and climates almost unimaginable to Earthlings. We travel there vicariously with the astronauts, who have brought back to the rest of us a new perspective on our world. To Earth-bound beings, the existence and relative proximity of the other planets provide the same potential respite and escape from our busy world that the ocean does for traffic-bound Bostonians or the mountains ringing Los Angeles do for those immersed in the smoggy basin below. Though we rarely travel to the oceans or mountains, we are comforted by knowing that we can. So too for the planets: they are real places to visit.

The planets are also a potential economic resource. Our ancestors transformed the wilderness—they harvested forests, harnessed rivers, and mined the earth. They built railroads through valleys, mountains, and desert floors. They farmed and settled rugged, virgin land. The economic and technological progress has made us (some of us) the wealthiest, healthiest, and most educated people on Earth. The solar system is also a wilderness to conquer, for good or ill. Although direct economic returns, such as mining the asteroids for metals, are not imminent, there are indirect benefits already. Studies of the atmospheres of Venus, Mars, and Jupiter have helped us understand our own so that we can better predict the weather. The need to fit sophisticated instruments into small spacecraft has spurred miniaturization technology. And, of course, the Space Program has provided employment to tens of thousands of workers.

What we know of the planets has been achieved by the research of scientists: astronomers, geologists, meteorologists, and others. What motivates them to do their research? Why are they fascinated by the planets? The planetologists who study the solar system share some of the same magical-mystery-tour and economic-political interests in space exploration that stimulate public interest. But their chief motivations are ones peculiar to their profession as scientists. Their innate curiosity about the universe is satisfied by their own special mix of rigorous methodology and hit-or-miss approaches to answering the questions and problems posed by what they see.

Planetologists have all of the psychological and sociological motivations that lead people to pursue their life's endeavors: recognition by colleagues, money to feed a family, or just resignation to following the path of least resistance through life. Some would-be scientists were told they were good in science in high school, followed their counselors' or parents' advice, took science curricula in college, and eventually emerged into the world equipped to be nothing but scientists, however their personal motivations might have evolved in the interim. It was unfortunate if society's educational system disgorged them into an economy surfeited with scientists, or if the inducements of government-financed scholarships led them to study space science, or particle physics, only for them to find at the end of their scholastic careers that society had now become interested in ecologists, cancer specialists, and exploration geophysicists.

But beyond pyschological and societal motivations, scientists have unique intellectual drives that differ even among the various disciplines. Perhaps what distinguishes scientists most from other educated men and women is their fascination with the solution of *simple* problems. It is a reflection of how poorly our schools teach science—and our culture disseminates it—that the common perception of scientists is that they deal with the most complicated and intricate problems, far beyond the ken of nonscientists. It may seem that scientific matters are beyond our everyday experience, yet human affairs are extraordinarily complex by comparison. It is particularly true that the public regards physicists' work as the most abstruse kind, while actually physicists study some of the simplest things in nature. For instance, the physicist who specializes in celestial mechanics despairs if the number of point-masses in his study approaches the number of fingers on one hand. The work of the geologist or biologist seems a little less remote, but actually the formation of a single valley, or the biology of a single organism, is far more complicated than anything studied by physicists. And sociologists, who study aggregates of the most complex animals on Earth, are the researchers most understood by laymen, although their questions might seem better left to the poets and philosophers.

The physical scientist (physicist, chemist, astronomer) is fascinated by a simple problem because there is a chance that something about it can be understood in a fundamental way. But few things in our everyday lives are sufficiently simple (physicists learned why a ball rolls downhill some while back). The physicist's studies may ultimately transform human society, but the physical entities he or she hopes to understand in minute and complete detail are frequently so microscopic and specific that there is little connection with familiar things. We all live because electrons surround atomic nuclei, but we can't hold either in our hands. As lay people we can vaguely empathize with biologists and cages of cute little mice. Many of us are ourselves amateur psychologists. But the world of the physicist's simple playthings seems remote, mysterious, and a little frightening.

Consider this ordering of intellectual disciplines: mathematics and physics, astronomy and chemistry, meteorology and oceanography, biology and geology, anthropology and psychology, economics and sociology, history, literature, and the arts. The motivations of the practitioners differ across this spectrum, as do the public's comprehension and the relevance we see for our personal lives and society. The list ranges from subjects that are simple to those that are complex; from thorough and fundamental understanding to partial and preliminary; from a highly logical to an intuitive and imaginative approach; from objects of study far removed from our everyday experience to our own environment and ourselves; from matters with only indirect social relevance to the very guts of our interpersonal relationships and our perceptions of the world.

As we traverse this intellectual spectrum we move also from a high degree of predictability to none at all: the physicist's prediction that water will flow downhill tomorrow never goes awry; the astronomer is pretty good at predicting when the sun will rise, although predictions of the splendor of Comet ("of-the-century") Kohoutek were a bit optimistic; the meteorologist's weather forecasts are a trifle less trustworthy; the prognostications of government economists are in the same league as the foretellings of palm readers; and the disciplines at the end of the list do not even claim predictive ability.

Still another criterion establishes the intellectual spectrum and is fundamental to the motivations of the practitioners: the information content of the data. It cannot be simplicity of subject that places astronomers, in whose purview are countless planets and stars, before the geologists, who study a single world. But the geologist is inundated with data: visual observations of hills, stream beds, fault scarps, rock beds, fossils, ripple marks, and—at still smaller scales—the texture, color, and fabric of the minerals that compose the rocks. The geologist can also measure the hardness of rocks, chemical composition, radioactivity, and so on. Such data are collected from locations around the world; the combined information is enormous and only the smallest portion of it can be digested. While the geologist can hardly understand the detailed history of a locale—every trickle of water, every vibration, and the crystallization of each mineral crystal—he or she can hope to decipher the largest and most important events. The geologist is thus one up on the social scientist; at least the subject is not constantly changing as it is being studied.

In contrast, the astronomer has far fewer data to work with. While other worlds may be as complex as Earth, the information that can be obtained about them is vastly less. Until this decade we could not touch, hammer at, or experiment with samples of the stars or larger planets. All we knew about them was what astronomers could see from a distant vantage point beneath our blurry atmosphere. (Modern technology does permit astronomers to "see" planets' radiation of infrared and radio wavelengths, beyond the colors seen by our eyes. Rockets and satellites send instruments above the atmospheric molecules that absorb and block the ultraviolet and other wavelengths.)

The other worlds and stars are so far away that they appear very small; most seem to be mere points of light even as magnified in the greatest telescopes. One can measure the brightness of a point of light at every spectral wavelength, how that brightness changes with time, and the position and subsequent motion of the point of light. That is absolutely all the data a ground-based astronomer can obtain from which to learn the nature of an object. Through intelligent interpretation and synthesis of such limited data from other similar points of light, the astronomer can apply scientific laws and other observations and can draw conclusions about the object that might seem wonderfully detailed to the nonastronomer. Although the astronomer will not approach the geologist's daily complexities, he or she will have learned absolutely everything possible about this object from the data on hand. If the astronomer designs a better instrument, uses a bigger telescope, or spends more time measuring the object through the telescope, a more accurate record may be obtained of the object's brightness.

But the data are always limited, so the astronomer extracts every possible clue, striving to reach that ultimate limit to his or her understanding of the object set by the meager means at his or her disposal.

An astronomer thrills at successfully solving the well-contained problem by rigorously stretching the few data to the limit. But a geologist is intrigued by the intellectual feat of recognizing the particular elements from an overwhelming mass of data that will help him or her understand a complex process just a little better. Each scientist has motivations, abilities, and intellectual habits suited to his or her own endeavor. Through the pluralistic efforts of all, the understanding of our environment—the universe— proceeds on many levels.

The simpler a problem, the closer we come to a fundamental solution. Conversely, the more we learn about a problem, the more complex it turns out to be, and the more difficult the solution is to attain. At times it almost seems that we are on a treadmill: better instruments yield more data; these raise discrepancies and complexities with respect to the simpler models that had been formulated from earlier data; and finally the data become so numerous that there is no hope of assimilating and understanding them all.

This problem is particularly acute in the planetary sciences. Until the last decade, our information about the major planets was obtained solely from telescopic observations of their small disks. Even the sharpest telescopic photographs of the planets were much inferior to a picture of someone's face taken with an inexpensive box camera. The fuzzy image of Mars through a telescope can be thought of as an array of several hundred separate points of light. By comparison, a television image, whether on your home set or a spacecraft vidicon picture, is an array of more than 100,000 separate picture elements. Recent technology has brought an explosion of planetary data: all nonredundant data obtained about Mars before 1965 could be printed in a small book, but Mariner

9, which orbited Mars for a year in 1971 and 1972, sent back more data to Earth *each hour* than had been collected previously about Mars in all human history. Volumes of Mariner 9 pictures alone form a small library, and this does not include the tapes of data returned by other Mariner 9 instruments.

The information explosion has been even greater in the case of the moon, which has been photographed and mapped in far greater detail than Mars. Moreover, the astronauts brought back half a ton of moon rocks that can be analyzed in microscopic detail for decades to come. It is said that a sufficiently large quantitative change is a qualitative change. So it is for planetology; accustomed to milking dry the limited available data, astronomers now feel inundated by spacecraft data and have turned the moon, Mars, and Mercury over to geologists. Several years ago an astronomer friend told me, "Mars is dead." He wasn't anticipating results of the Viking Mars-lander biology experiments, which searched for life on Mars; he meant that there was nothing further that astronomers could learn about Mars. Astronomers are also starting to turn Venus and Jupiter over to meteorologists now that vast amounts of atmospheric data have been returned by Mariner and Pioneer missions to those planets.

This scientific version of musical chairs is evident in the programs of the annual meetings of the Division for Planetary Sciences of the American Astronomical Society (AAS/DPS). It is the organization of the couple of hundred astronomers whose major interest is studying the planets and solar system. In 1971 and 1972, 36 percent of the scientific talks presented at the meeting dealt with the moon and Mars. By 1976 only two talks dealt with the moon and fewer than 9 percent with Mars. Meanwhile, papers concerning the asteroids and the natural satellites of the planets increased from 9 percent of the program in 1971 to 37 percent by 1976. Why satellites and asteroids? They are barely more than points of light and, except for the moons of Mars, have yet to be examined closely by any spacecraft. What could be a more ideal pursuit for a planetary astronomer than to turn a telescope toward those small, faraway worlds and discover all that can be learned about them?

Planetary scientists are guided, beyond their intellectual preferences, by spiritual and experiential motivations. Although astronomers spend most of their time in laboratories or classrooms, they make periodic pilgrimages to mountaintop observatories. Beneath these cold, distant pinnacles on the doorstep of the heavens are spread the Earth's desert wildernesses. Civilization is manifested only by the headlights of an infrequent automobile and the glowing auras of cities on the far horizon. Amid the towering pines an occasional dome is silhouetted against the stars and the Milky Way. Inside, the giant machinery looms skyward, slowly revolving to follow the stars. The astronomer sits below, monitoring the instruments, checking the guiding, and gazing at a slice of the heavens through the open slit in the dome. Drowsily intoxicated by the rarefied air and the shock of the body's circadian rhythm adjusting to nighttime wakefulness, the astronomer is receptive to the immanence of Nature.

I recall once at the National Observatory; the night was still and silent save for a few howling coyotes. I was trying to measure the light from some faint asteroids, a task that grew more difficult as the full moon rose to dominate the sky. But sitting on the carpeted floor looking out the slit, I noticed the moon growing dimmer. Its left side began to be eaten away by (I then realized) the shadow of the Earth. Finally it became a faintly glowing ocher orb, and the scattered moonlight in my photometric signal vanished entirely. Between swallowing my midnight lunch and marveling at what was to me a quite unexpected total lunar eclipse, I hurriedly measured several asteroids right next to the ghostly moon in the sky. As dawn approached, the moon emerged from the Earth's shadow and grew full again. As it sank toward the west, I was finishing measurements in the brightening eastern sky. As the sun's first rays peeked over the Rincon Mountains a hundred miles to the east, I trudged down from the dome to sleep as the rest of the world awakened. I knew that the magnetic tape was now filled with good data on tiny worlds never before measured. I was overcome by a profound sense of satisfaction and I felt as renewed as the moon had been earlier that morning.

A geologist laces up thick-soled boots, packs away a compass and trusty rockhammer, rents a four-wheel-drive Jeep, and heads off to wildernesses as strange as those on other worlds. The exhilaration is tangible as the geologist shades the sun burning down from cloudless skies and climbs up a barchan sand dune in a desert that hasn't seen a drop of rain in over a century. Another geologist clambers over the plateaus and valleys of frigid Antarctica, where liquid water, which nourishes most of the world, hasn't existed for millennia. Perhaps this remote spot is more like the surface of Mars than anywhere else on our variegated planet.

Still other geologists seek planetary analogs in rare places where the Earth is being born. They step gingerly across glassy, solidified lava that was molten but a week before. Their next step might pierce a fragile bubble and plunge them into a fiery cave below. The mists and sulfuric fumes swirl over the unearthly new landscape, threatening to suffocate the explorers; yet they know that the deadly gases are the Earth's freshest emanations and they are reminded of how "unnatural" is the oxygen-rich air we breathe. The intrepid explorers watch a glowing pink tongue of molten rock ooze past them down a glassy channel. Tired and hungry, they stop to bake a frozen pizza on a freshly crusted lake of liquid rock. But the molten pudding shifts, carrying the simmering meal a few feet out "to sea." They plunge onward still hungry, for one does not dip one's toes into a puddle a thousand degrees hot. At last they stare down into the fiery hellhole they have journeyed to see—a caldera ¼ mile across. Silvery-orange liquid rock from the Earth's deep interior swirls in turbulent, incessant motion. At any moment the precipitous cliff from which they observe in awe may break loose and crash into the all-consuming liquid furnace below. Undaunted and dedicated, they set up their equipment before retreating to relative safety. Volcanologists' lives are as fraught with danger as those of any explorers of centuries past; and from their measurements and observations some further understanding may be gleaned of processes that shaped the Earth, and Mars, and other worlds too, eons ago.

The geologist in the wilderness epitomizes man against nature. A tiny, transitory creature, whose life will be over in a mere second of geologic time, attempts to discern the life processes of Earth, whose metabolic rate is a million times slower than his own. Hammering at the rocks pregnant with data, the geologist is in physical contact and combat with the Earth, like a mosquito attacking Goliath. Ever so grudgingly, the Earth yields its life history.

In a windowless, air-conditioned room in the basement of a university building, another planetary scientist finishes preparing her experiment. She was trained as a chemist but now calls herself a meteoriticist. The elaborate machine before her cost over $100,000 and incorporates the latest advances of American engineering technology. For the past many months she has been fabricating auxiliary equipment and calibrating the instrument so that she can be sure the readings will be accurate. Now she has dialed in the proper numbers and is ready for her first measurement.

She picks up a stone that has been lying in a nearby tray, her hand trembling ever so slightly. This small lump of rock is older than any material known to mankind—billions of years. For the last many millions of years it has traveled through space to be delivered to Earth, and then to her laboratory. She returns it to the tray and picks up instead a small plate she constructed several days earlier; it contains a thin slice, only a few hundredths of a millimeter thick, from the same meteorite. She puts the plate in the sample chamber of the instrument, pushes a couple of buttons, and watches expectantly as the preprogrammed instructions are carried out. A fine beam of electrons plays across the thin section;

from the reemitted X-rays, the precise chemical composition of each tiny spot on the slide will be measured.

Though she has never looked through a telescope or mapped a geological quadrangle, the meteoriticist regards herself as descended from the purest planetological tradition. For decades, even centuries, before the astronauts returned the first moon rocks, her colleagues and their progenitors had been studying rocks that came from planets a hundred times farther away than the moon. After all, it was the Nobel Prize–winning chemist Harold Urey who had written the definitive book on the planets long before modern astronomers thought our neighboring worlds worth studying. Soon the meteoriticist will have the microscopic chemical analyses that will permit her to test her theory about the processes that assembled this rock, grain by grain, an unfathomable eternity ago.

2

Craters:
Planetary Chronometers

ABOUT 25,000 YEARS AGO a small band of native American hunters slowly made its way among the sparsely scattered pines then growing at the southern edge of the mesas that have now been inhabited for a few hundred years by the Hopi Indians. The hunters were descended from nomads who had trekked across the land bridge from Asia to exploit the riches of the vast continent. Spurred by the knowledge that they needed meat within a few days, the men were single-mindedly tracking the nearest prey.

A sudden flash of light in the sky above startled them. They turned to witness a brilliant ball of fire streaking toward the plains to the south. It seemed to rival the sun in splendor, and in a few short seconds it reached the horizon in a final blinding explosion of light. The fireworks were snuffed from view a few seconds later by a columnar pall of dust that lifted upward and outward from the point of impact.

Temporarily forgetting their quest for food, the men stood

oblivious to the twittering birds nearby, gazing in amazement as the dust cloud rose ever more slowly upward from the horizon. There had been legends of fireworks and of spectacular clouds emanating from the black volcanic hills to the southwest, but this lightning bolt from an otherwise cloudless sky was something they had never heard about, let alone witnessed, before.

Without warning the ground jolted and shook violently, knocking the hunters from their feet. Then the earthquake quickly subsided. But as the men struggled back to their feet they saw a wall of dusty haze expanding toward them across the plains below. They began to run away from the edge of the mesa. But before they had gotten far, their still world was filled with a roaring, rumbling noise. Terrified, they dropped again to the ground just as a mighty concussion of wind swept over them. They shut their eyes to the swirling dust. The wind died as abruptly as it had begun and the hunters stood up once again, glancing fearfully toward the south and wondering what further surprises were in store for them. But little was to be seen through the haze. The hunters anxiously queried each other about the intentions of their gods in bringing these calamities upon them in such sudden succession. They debated about whether to call off the hunt for the day, but hunger compelled them to press on. Within an hour, their world was back to normal again, except for some lingering haze, and stalking the animals was once again uppermost in their minds.

What the members of this hunting party could never have realized was that they had witnessed one of the most spectacular and unusual of the geological processes that shape the face of Earth. Since the prehistoric formation of the Arizona Meteor Crater just described, no other human beings have witnessed the nearly instantaneous creation of such a large landform. Meteor Crater, about 1 mile across, was formed by the impact, at a speed of more than 10 miles per second, of a huge nickel-iron meteorite or asteroid fragment about 100 feet across.

Rare though such cratering events are in human experience, impact cratering is actually the dominant geological process in the solar system. The "mountains" on the moon, Mercury, and Mars are mainly the raised rims of old craters. The same is probably true of Venus, the asteroids, and many of the satellites of the outer planets. Indeed, cratering may have been the dominant geological process on our own planet during its first half-billion years of existence, just as continental drift, running water, sedimentation, and volcanism have dominated terrestrial geology in the most recent half-billion years during which plant and animal life have proliferated on Earth.

As we compare our own countryside with the moonlike cratered surfaces of most of the other rocky worlds in the solar system, we might ask why there are so few large craters on Earth that even a relatively tiny one only 1 mile across is a famous landmark. On the other hand, if interplanetary space is so empty, how can there be so many scars from impact explosions on most planetary surfaces?

The answers to both questions involve time. To be sure, interplanetary space is very empty indeed. If we were to reduce the scale of the inner solar system (those planets closer to the sun than Jupiter) to fit inside Houston's Astrodome, the Earth would be the size of a pea and there would be scattered throughout the rest of the stadium a few dozen microscopic dust particles capable of producing craters on our Earth-pea similar in scale to typical lunar craters, such as those shown on page 52. The Earth-pea and few dozen dust grains would be moving along their orbits in the Astrodome at an imperceptibly slow speed, similar to the speed of a watch's minute hand. Space is very empty, and the chances for collisions might seem nil. But after $4\frac{1}{2}$ billion years, which studies of meteorites have shown is the age of the solar system, our Earth-pea would have traveled 2 billion miles in the Astrodome, and each dust grain a similar distance. After having covered such great distances within the confined volume of the Astrodome, the Earth-pea probably would have struck most if not all of the dust grains!

The Earth, then, probably has encountered most of the projectiles that have crossed its orbit in space. Most of those that failed to strike the Earth have scarred the surfaces of the moon and other planets. There are a lucky few asteroids in the inner solar system that by chance have missed the planets so far, while one by one their siblings died explosive deaths. Other asteroids now nearby were originally in safer orbits that did not cross the orbits of the larger planets, but they have shifted recently into Earth-crossing paths.

Where are all the craters that must have been formed on the Earth? Why don't we see them? Perhaps we do see a few of them, if we look carefully enough. Some geologists speculate that Hudson's Bay in Canada, or at least part of it, may be the flattened and flooded remnant of a huge asteroidal impact crater. The remains of several other probable craters, up to tens of miles across, have been discerned from aerial photographs by the trained eyes of photogeologists. The extremely eroded condition of those several large craters that have been found suggests why the many others are completely absent: compared with the immense durations between major impacts on Earth, the geological processes that deform and erode landforms proceed exceedingly rapidly.

During our own lifetime the modification of the landscape by floods, glaciers, and earthquakes has been very slight indeed. And the major forces in the Earth's crust that cause the continents to drift apart and crash together, creating whole mountain chains, occur even more slowly. Yet they occur rapidly enough to have erased all evidence of large terrestrial craters except in those ancient continental cores called pre-Cambrian shields. Most of the couple of hundred terrestrial craters that have been discovered are located in shields, such as the one that comprises much of Canada and some of the north central United States. Such regions have been protected from the mountain building, flooding, volcanism, and other destructive geological activity common along continental margins.

Meteor Crater itself is already appreciably eroded, although it was formed only a few tens of thousands of years ago. Although it is still impressive, dozens of arroyos and channels have been carved down its inside slopes by the raging run-offs of hundreds of thousands of thunderstorms. In a million years there may be little or no trace of Meteor Crater; yet a million years is only a few ten-thousandths of the age of the Earth. Given the near eternity of geological time, the crustal processes active on the Earth today can wipe the landscape virtually clean of craters.

Instead of asking why the Earth today has so few craters, we should wonder that the moon and other planets have retained so many. The answer is simple: other planets must lack the powerful erosive processes active on our world. (Logic forces a second alternative upon us: perhaps the other planets share the Earth's geological activity but are being struck by thousands of times as many asteroids as the Earth. Although we have no assurance that the bombardment is the same on all planets, it would take a cosmic marksman of exceeding skill to single out the moon and other planets for bombardment but avoid hitting the Earth.)

One of the major conclusions of modern planetology is that our planet is, compared with the other inner planets, a geologically active world. The familiar mountains, valleys, hills, and coastal plains have resulted from processes special to the Earth. For decades we had viewed our world as a typical planet orbiting a typical star. In 1965 scientists and laymen alike were shocked when Mariner 4 revealed a moonlike landscape on the planet Mars rather than the mountains and valleys to which we are accustomed. But as Mariner 10 has shown for Mercury, moonlike landscapes are the norm in the solar system. And so our perspective on our own world is being changed. It may not be entirely coincidental that the scientific revolution taking place in geology, epitomized by new conceptions of continental drift and plate tectonics (see Chapter 9), is happening simultaneously with the exploration of the other planets.

Craters or pits are conceptually and geometrically simple. Indeed they are common, from the tops of frying pancakes to the old battlefields of Indochina. Whenever a sudden, outwardly radial force is exerted on a horizontal surface in a gravitational field, a crater is likely to result. Terrestrial volcanic explosion craters (called maars) bear a striking resemblance to impact craters, and debate long raged over whether the large lunar craters were of volcanic or impact origin. Detailed studies of terrestrial craters have revealed subtle diagnostic differences, and the debate on lunar crater origin is now largely resolved in favor of impact. But the major features of a crater (its size, depth, and distribution of ejected material) can be predicted simply from the energy imparted to the ground at a point. It matters little whether it is the kinetic energy due to a meteorite's mass and velocity or the explosive energy of a bomb: the same size crater results each time from the sudden release of the same energy. In an instant, the explosive energy liquefies and vaporizes part of the rock and imparts a violent shock that expands into the ground, tearing rocks apart and heaving them upward and away. The larger, slower-moving fragments build up on the exterior rim while high-velocity fragments land at great distances from the crater.

Impact cratering experiments have been done at NASA's Ames Research Center near Palo Alto, California, using hypervelocity bullets traveling many miles per second. NASA experimenters have learned how crater size and form change with different velocities, bullet masses, angles of impact, and surface material strengths. Scientists also have a much better idea now of the interplanetary population of asteroids, comets, and smaller projectiles that have created craters in the past and still occasionally impact on planets. Using the 48-inch Schmidt camera-telescope on Mount Palomar, astronomers have surveyed the fainter asteroids, especially those passing relatively close to the Earth and Mars. Theoreticians have calculated the permanence and longevity of these projectiles and the chances of their impacting a planet or being moved by the gravity of other planets into more distant orbits that cannot intersect the planets.

Thus scientists have a firm grasp on all major aspects of the cratering process. But understanding the cratering process is just the beginning, not the end, of planetologists' interests in craters. By applying this knowledge, scientists have been able to decipher the relative geological histories of the moon, Mars, and Mercury in the absence of fossils, which in the last century were necessary for establishing the Earth's chronology.

Little was known about the geological history of the Earth until the early 1800s, when Charles Lyell and others noticed that different groups of fossils of extinct species occurred in an orderly sequence in deeper and deeper layers of stratified rocks. Although the concept of biological evolution was yet to be developed, geologists pragmatically recognized that if a layer of limestone contained certain ammonoid fossils, it must have formed during the Devonian Age (now known to have been about 350 million years ago), even though in all other respects that limestone was indistinguishable from limestones formed before and since. By tracing the different fossil distributions in rock outcrops throughout the countryside, European geologists developed a stratigraphic sequence of rock units ranging from ancient to modern. Later, fossil groups and rock units were correlated around the world and a coherent sequence of rock-forming and erosive periods was synthesized for the entire Earth.

A relative sequence of geological events provides no clue to their absolute ages. To say that coal deposits formed in Pennsylvania at the same time coal formed in England, during the Permian Age, does not distinguish between the notion that the Permian occurred during one tropical summer in 2350 B.C. and our present belief that it was about 250 million years ago. The relative history of terrestrial geology has been converted to an absolute one by the techniques of geochronology, perfected in the middle of this century. Certain radioactive elements found in rocks are known from physics to decay, in known durations, into different elements of special isotopic composition. From measurements of such isotopic components in a rock, its absolute age may be accurately determined. The oldest rocks dated on the Earth were formed about

3.8 billion years ago, which is a common age for most lunar rocks. The same geochronological techniques applied to meteorites generally show them to be about 4.6 billion years old.

For Mars, Mercury, and the other planets we do not have rock samples to date. Nor can we even establish relative sequences from fossils. Here is where craters come to the rescue. But let me begin with an analogy.

Consider a postman (of the old-fashioned variety) walking down the sidewalk from house to house, delivering mail. Or shall I say trying to walk down the sidewalk, for let us imagine that it has been snowing for two days. It is not a heavy blizzard, but the snow has been steadily mounting up and is a foot and a half deep on the undisturbed lawns. The postman is no stratigraphic geologist, but he can conclude when, in the past couple of days, the different families were out shoveling the snow from their walks. The snow is a foot and a half deep in front of the first house and no trace of the sidewalk is evident; either these people are on vacation, are sick, or are just plain lazy, our postman concludes, for clearly nobody has pushed a shovel across this walk for two days. In front of the next house the snow is only three inches deep—evidence that the walk was shoveled earlier that morning. Apparently the next sidewalk was last cleared the previous day, since about a foot of snow has fallen on it since. At the last house on the block, the sidewalk is practically bare, with only a few scattered snowflakes sticking on the cold cement; the postman now notices Mr. Smith still at work, finishing his driveway.

Substitute *lunar surface* for *sidewalk* and *cratering projectile* for *snowflake* and you understand the basis for the lunar relative stratigraphic sequence, first derived in the early 1960s by scientists at the Astrogeology Branch of the United States Geological Survey. While all lunar regions are more heavily cratered than anywhere on Earth, some are more cratered than others. This is an important observation, for it implies (reasonably assuming that interplanetary projectiles strike all sides of the moon equally) that the moon has not been a geologically dead body, just passively recording the scar of each impact. Rather, on some parts of the moon, the craters once formed have been covered up or otherwise erased, and only the more recent ones survive.

The provinces on the moon most devoid of large craters are the maria, or lunar "seas," the dark, circular patches visible with the unaided eye from Earth. Compared with the crater-upon-crater zones on the brighter uplands of the moon, the maria have only about $\frac{1}{20}$ as many large craters. Even among maria provinces, crater densities differ by up to a factor of 4. If we could be assured that the impact rate has been constant, we might conclude that the maria were formed only $\frac{1}{20}$ of the moon's age ago, or about 200 million years ago, when dinosaurs were roaming the Earth. But unlike our postman who witnessed the two days of steady snowfall prior to making his rounds, we cannot presume that the interplanetary bombardment was steady. More likely there was an early "blizzard," a phenomenon that would invalidate the 200-million-year-age estimate.

Observations other than of crater density can be made from lunar photographs, taken both through telescopes and from spacecraft, which clarify relative sequences on the moon. For instance, specialists in photointerpretation quickly recognized the scalloped feature in a photograph as the edge of a huge once-molten lava flow that spread across the mare surface, obliterating all preexisting craters. One need not count craters to conclude that the land underneath was formed before this latest flow lapped upon it! Only in small localities do such geometrical superposition relationships reveal relative age sequences. Crater count comparisons are required to establish the ages of these flow units relative to features elsewhere on the moon.

Applying such procedures to the best lunar photographs, geologists have devised a relative sequence of lunar geological history that is as reliable as that for the Earth. Craters, then, have served admirably in place of fossils. Yet before the Apollo 11 astronauts brought back lunar rocks in 1969, we had virtually no *absolute* age calibration for the moon. Still the picture wasn't totally bleak. In the mid-1960s Arizona astronomer William Hartmann, among others, argued that the maria had to be much older than 200 million years. He estimated the rate at which the Earth, and thus presumably the moon as well, is bombarded by interplanetary debris. He used counts of the nearby asteroids, of the meterorites that fall each year, of the meteors that flash through the night sky (meteoroids too small to make it to the ground), and of the comets that pass by. By estimating the sizes of these projectiles, and how large the craters they might create would be, Hartmann showed that the present cratering rate over $4\frac{1}{2}$ billion years would create only about the number of craters visible on the lunar maria, not the twenty-times-greater number on the highlands. Hartmann concluded that there must have been a blizzard of bombardment sometime, and he suggested that it most reasonably happened near the beginning of lunar history. Hence, long before the Apollo landings most lunar specialists believed the maria to be $3\frac{1}{2}$ to 4 billion years old, and that the highland craters formed earlier, during the final stages of formation of the moon itself.

Shortly before the landing of Apollo 11 some geochronologists, disinclined to believe that ages could be established by any method other than their own, dismissed the validity of the crater count method of determining ages. To make matters more confused, one influential astrogeologist then obtained some classified military data that showed (to the privileged few with security clearances) that many more objects were entering the Earth's upper atmosphere than had been estimated from meteor observations. But apparently these military data suffered from exaggeration, for when the geochronologists measured the ages of Apollo 11 lava basalts from sparsely cratered Mare Tranquillitatis, they turned out to have formed about $3\frac{1}{2}$ billion years ago after all! The terrestrial bombardment rate today is admittedly still somewhat uncertain and could be higher than the average over the last few billion years. The whole question of bombardment rates is an important area of active current research.

I have described how the unprejudiced manner in which cratering impacts strike all parts of a planet equally enables scientists to develop relative geological histories for the planets. Beyond that, the remarkable similarity of the initial shapes of craters of all sizes enables planetologists to study the nature and styles of erosion and landform degradation on other planets.

The branch of geology known as geomorphology is concerned primarily with the processes that carve canyons, erode mountains, and otherwise shape our landscapes. Running water is the chief erosive agent on the Earth, although the chemical weathering of rocks, glacier movement, and many other forces help to shape it. The moon is a simpler place, since water is absent and the sole erosive agent is the abrasion of surface rocks by impacting meteroroids. In times past, the destructive impacts of asteroids and the flooding of basins by molten lava played major roles, but those were the only forces.

Mars is a more complex world than the moon, although today it too lacks running water. Unlike the moon, Mars has a thin atmosphere, and the photographic record from the Mariner 9 spacecraft revealed abundant evidence of the abrasion and filling of Martian landforms by windblown dust and sand. In past ages, lava flooding and cratering changed the face of Mars, but so did running water and other processes. The shapes and forms of Martian craters, as they have been gradually obliterated over the eons, reveal much about these forces that have shaped the surface of the Red Planet.

To oversimplify a bit, imagine a dusty Mars where craters of all sizes are being formed regularly, while dust is continually settling down and gradually burying them. (Other erosive processes, singly or in combination, degrade craters in roughly similar ways, but dust deposition is easy to visualize and is certainly one of the processes currently active on Mars.) Each crater is initially formed as a fresh, bowl-shaped excavation like Meteor Crater in Arizona. But as the dust settles down, the crater bowl becomes shallower and shallower until it is finally filled in. Of course, small craters are buried much more rapidly than larger ones, but there are always new small craters being formed while others created before are only partly gone. So at any time there is a complete range of craters of all shapes, but only up to a certain size.

Suppose the total depth of dust deposited on Mars since its surface formed is 1 mile. Then many generations of small craters would have been covered and only the most recent ones would still be there. But craters larger than 20 miles across (with depths of 1 to 2 miles) would still be visible; even those formed very early in the history of Mars would not yet be completely buried. Thus, as the Mariner 4 spacecraft first revealed in 1965, the proportion of small to large craters on Mars shows a lack of small craters compared with the proportion of small to large lunar craters or small to large asteroids. The Estonian astronomer Ernst Öpik, who has resided for some decades in Northern Ireland, first correctly ascribed the absence of small craters on Mars to a major erosive process on that planet. (Öpik has always been one of the most resourceful planetary physicists. He still recounts the tale of his departure from Tashkent during the Bolshevik Revolution of 1917. The train had to stop at little stations for hours and even days as local stationmasters of uncertain loyalties debated what to do. But on more than one occasion Öpik pointed to the red or white glow of the aurora borealis in the northern skies, suggested it was the light of the appropriate advancing army, and got the train rolling again!)

Consider the shapes of Martian craters. They range from fresh craters to highly degraded ones. In the example described above, both cratering and dust deposition occur continuously, so for all sizes there is the same proportion of craters in each stage of degradation. There is an analogy with age distributions of human beings: we could classify people as children, young adults, middle-aged, and elderly. Although children are always being born and the elderly dying, the percentage of people in each of the four age groups remains constant, at least in a stable society.

Societies, however, are rarely stable. Age statistics of Americans reveal a disproportionate number of young adults (especially those in their late twenties). Even if we knew no history at all, we might infer from this demographic fact that there was a baby boom at the time these young adults were born. Rather analogous studies of the proportions of Martian craters of different classes reveal anomalies compared with the stable dust-filling example previously described. Ken Jones of Brown University and I have concluded that there was an episode in Martian history when the crater obliteration rate greatly exceeded the cratering rate in comparison with more recent times.

It would be too simple to ascribe this episode of crater filling to dust alone. But whatever combination of processes was active during this episode, it had a major effect that is still visible on the Martian landscape. Moreover, the relative stratigraphic sequence for Mars that has been established, using the same principles I described earlier for the moon, shows that this period of rapid crater filling coincided with an early major volcanic period on Mars and with the period when water apparently was flowing in channels throughout the equatorial regions of the planet. This fascinating period of Martian history occurred after the major cratering bombardment had slackened, but long before the greatest Martian volcanoes reached their present stature. Exactly when this age of Martian renewal took place we cannot say until we have some Martian rocks to date or until we can establish the Martian cratering rate much more precisely than we have so far.

From the simultaneity of the formation of the volcanic plains, the creation of apparent river channels, and the crater obliteration episode, it is tempting to imagine that Mars was a much more Earth-like planet for a time. Perhaps rising temperatures due to radioactive decay in the Martian interior caused volcanism and at the same time melted and released into the atmosphere great amounts of previously frozen water from the Martian soil. There might have been great wind and rainstorms in the temporarily Earth-like atmosphere that filled the craters and created the river valleys we see today. Later, most of the atmosphere was lost to space or was frozen out at the poles; the rains ended and Mars became the dry, dusty, practically airless world it is today. It is an enticing scenario that has been devised from analysis of Martian craters, but it is only a tentative one that requires confirmation from further studies of the Red Planet.

3

Uniformitarianism and Catastrophism

> And the waters prevailed exceedingly upon the Earth; and all the high mountains under the whole Heaven were covered.
>
> —GENESIS 7:19

A WILD ANIMAL is alarmed by instability in its ecological niche, a potential threat to its life. A sufficiently widespread calamity may threaten an entire species. Mankind has advanced by modifying the environment and so far has managed, more or less, to adapt. But the accelerating changes of the twentieth century have been psychologically disturbing—and for good reason. We discover that we have introduced carcinogens into our food, water, and air, while pollutants and aerosol sprays threaten to plunge us into a sudden climatic change, perhaps a new ice age. Some people would promote the SST, continue dumping asbestos into Lake Superior, and relax pollution standards for the sake of our immediate economic health. They believe that we can adapt limitlessly and that science and technology can rescue us from potential catastrophe. But it is a faith in technology that our instincts tell us may be poorly placed. Hence, an alternate impetus toward leading simpler, more natural lives.

26

The optimum environment of our dreams is the Garden of Eden or a tropical paradise, where the landscape is natural and fixed, the climate mild, and seasonal changes slight. The real world, with its storms, earthquakes, and floods, is not so constant. But as contrasted with our twentieth-century "improvements," we conceive of Nature as having been fairly constant over the millions of years during which mankind evolved. We feel comfortable with our "uniformitarian" conception of Nature's constancy and we fear change, especially catastrophic change.

Those who believed in an active Divinity and the wretched state of mankind and who looked to salvation in the next world were often inclined to believe in catastrophic changes in this world. No catastrophe was beyond the power of the Old Testament God to bring against his wayward children. Yet he might also save men, as he saved Noah from the Deluge. For one whose fate was in God's hands, no constancy of environment was required for salvation. Catastrophism is still proclaimed in "The World Tomorrow" broadcasts of the Radio Church of God, which predicted as late as the 1960s that unprecedented catastrophes would culminate in the early 1970s, prior to divine salvation of the chosen few.

Since the uniformitarian/catastrophic dialectic is so fundamentally rooted in our psychology, culture, and religion, it is not surprising that the science most concerned with the natural environment—historical geology—has been greatly affected by philosophical and religious beliefs about environmental stability. Even debates concerning the geology of other planets are being waged today on the same philosophical battleground.

Most modern scientists assume the constancy of the laws of physics. The Book of Genesis, literally interpreted, and the theories of the catastrophist Immanuel Velikovsky are in conflict with even that assumption. But the uniformitarianism on which modern geology was founded nearly two centuries ago as a reaction to biblical catastrophism is a much more sweeping but less secure assumption. Increasingly, strict uniformitarianism seems incompatible with the scientific evidence, although modern geology texts still rely on it and many geologists are still in its grips. It may well be the unfolding history of the solar system inferred from space exploration that will ultimately synthesize a reasonable blend of uniformitarianism and catastrophism, free from religious or philosophical bias.

I need not dwell on biblical catastrophism. If there were any truth to Archbishop Ussher's calculation of the date of Creation as 4004 B.C., the geological processes we witness today could hardly have carved deep canyons, laid down thousands of feet of sediments, and deposited aquatic fossils a thousand miles from the nearest sea. Even today, the visitor to the Grand Canyon is hard pressed to accept geologists' contentions that the river carved the whole chasm in only millions of years. By the early 1800s the Deluge of the Old Testament was deemed inadequate to explain European geography and mass extinctions apparent in the fossil record. So natural philosophers invoked a whole sequence of catastrophes, supplemented by divine repopulations of the world by ever more advanced animals.

Uniformitarianism, enunciated by James Hutton in the late eighteenth century, emerged from the Age of Reason, although its conceptual foundation can be traced from the medieval church and even the Greeks. It is the doctrine that "the present is the key to the past," that the Earth may be understood by assuming that the natural forces acting today have always been acting similarly in the past. Great canyons result from rocks being washed downstream grain by grain, and rock layers are built from sedimentation, grain by grain. Released from the 4004 B.C. constraint, the history of the Earth stretched back into the indefinite, unchanging past.

But catastrophism did not die easily, and by the early nineteenth century geologists had split into ideological camps. Uniformitarian geologists, led by Charles Lyell, hardened their hypothesis into dogma. It then took a while to gain their acceptance of even such limited catastrophes as ice ages. By the twentieth century it was axiomatic that the fixed continents were in perpetual balance between the steady forces that uplift and create new land and those that gradually wear it down. That Africa might once have been adjacent to South America was regarded by the prevailing uniformitarians as being as outlandish as the notion that craters on the Earth and moon were created by catastrophic collisions.

A methodological assumption akin to uniformitarianism pervades science: among alternate hypotheses, the simplest and most straightforward is to be preferred. If the Earth as we see it can be explained by present processes, why invoke catastrophes or changes? Of course, changes or catastrophes are not thereby disproven, and should contradictory evidence subsequently appear, a more adequate theory must be sought.

However, geologists read more into uniformitarianism than methodological simplicity. They increasingly felt that environmental constancy was a proven fact and that geological processes necessarily act within narrow limits. Rather than abandoning a simple model, geologists went to great efforts to explain contradictory evidence in uniformitarian terms. More than a century ago, William Whewell attacked Lyell's uniformitarianism, asserting, "Whether the causes of change do act uniformly;—whether they oscillate only within narrow limits;—whether their intensity in former times was nearly the same as it now is;—these are precisely the questions which we wish Science to answer. . . ." But geologists rarely cared to investigate whether the Earth might have been different in the past or to determine the degree to which geological processes have been nonuniform. After all, the whole superstructure of their science was built on uniformitarian assumptions; to ask a geologist to question them would be like asking a priest to be skeptical of God or an army general to question the utility of war.

The processes that shape our world are far from constant. From gentle breezes emerge tornadoes, packing winds of many hundreds of miles per hour. A mountain of cinders once grew in a Mexican cornfield. The largest lake in California, the Salton Sea, was formed practically overnight by diversion of the Colorado River. In 1883 most of the island of Krakatoa vanished in an explosion that threw about 20 cubic kilometers* (5 cubic miles) of rock into the sky.

Though acknowledging that such minor catastrophes as floods, storms, and landslides "should not be discounted," an author of a recent geology text nonetheless concludes, "The cumulative effects of the constant downslope creep of soil, the gradual decay of rocks under the atmosphere, and the gradual removal of material by streams hour after hour, day in and day out, over millions of years are probably much greater than the effects of these minor catastrophes."

Recent research contradicts such a uniformitarian view, showing that the greatest changes occur during the biggest storms, earthquakes, or other "minor" catastrophes that take place during any interval. For instance, Eugene Shoemaker of the California Institute of Technology and the U.S. Geological Survey has shown that most change in the Grand Canyon results from severe storms and floods, not from the gradual action of the river and ordinary rains. Shoemaker also, in his doctoral dissertation two decades ago, proved convincingly the catastrophic mode of formation of Meteor Crater; he subsequently became the leading lunar geologist of the 1960s.

Modern research has also shown, against uniformitarian resistance, that still more powerful catastrophes occurred in the past. A vast region of valleys, basins, and buttes, 200 miles on a side, was created in eastern Washington State in just a few days 18,000 years ago, when 400 cubic miles of water from Lake Missoula suddenly broke through a glacial dam and poured across the land to the Pacific. There are indications that ice ages may begin almost instantaneously on a geological time scale, well within a human lifetime, so Thornton Wilder's image of the glacier advancing on

---

*A kilometer is a bit more than ½ mile, a centimeter a bit less than ½ inch. I use metric and English units interchangeably in this book. We must soon get used to metric units, but it is hard to switch instantaneously.

George Antrobus's house may be nearer the truth than the allegory of all human history that Wilder intended in *The Skin of Our Teeth*. Great evolutionary changes have occurred also. For instance, about a hundred million years ago, the whole Indian subcontinent broke off from east Africa and crashed into southern Asia, creating the Himalayas. Although continents drift only inches per year, the idea that continents move at all was beyond the pale for uniformitarian geologists during the first two-thirds of the twentieth century (see Chapter 9). Other dramatic ways in which the planet Earth has evolved are still being uncovered; it is now also thought, for example, that our atmosphere was very different in the past.

It is most reasonable to constrain geological uniformity only by the laws of physics. Given sufficient energy, we are limited only by our imaginations in conceiving plausible ways in which the energy may be organized into geological activity. In making an inventory of accessible energy, we are ultimately limited by Einstein's $E = mc^2$, though such perfectly efficient conversion of mass to energy is not generally realizable. Until the present century the only known energy sources were sunlight (rather feeble for moving continents!) and the residual heat from creation (planets were assumed to be formed from molten rock). Lord Kelvin had shown a century ago that any primordial heat would be radiated away in 20 million years, so the apparent lack of adequate energy for propelling continents became an early obstacle to acceptance of drift.

We now believe the major source of energy driving the Earth's active geology is heat released in the interior by radioactive decay of uranium, potassium, and thorium. Other sources of heat that may have been important when the Earth was young include gravitational (tidal) interaction among planets, explosive impacts and collisions, energy released when heavy metals plummet to form a central core, interaction of a planet with the solar wind of charged particles, and the decay of very short-lived radioactive elements.

While Earth scientists understand the generation of heat

within our planet, it is much more difficult for them to understand how that energy is converted into the organized motions of the drifting continents and spreading sea floors that build mountains and shape our world. The complex suite of geological processes, belatedly recognized by geologists and dubbed plate tectonics, is awe-inspiring. It is much more intricate than the comparatively simple solidification and contraction of Mercury around its immense iron core, which has caused its crust to buckle, forming the large wrinkled cliffs photographed by Mariner 10. Our own planet's mantle convects turbulently, similar to the rising and sinking of air near a thunderstorm or water in a pan approaching a boil. Our drifting continents are carried on the backs of the churning, subterranean currents, and molten rocks exude from cracks in the ocean floors. We will examine later how this fundamental internal life of our planet is ultimately responsible for the life-sustaining environments on the Earth's surface. How all this dynamic activity is produced is as yet poorly understood and awaits more detailed comparisons with the other rocky worlds.

The laws of physics also permit such unexpectedly complex, organized motions as weather fronts, jet streams, and cyclones that are driven in our atmosphere by the energy difference of absorbed solar heat between the tropics and the poles. Still other planetary processes of almost unimaginable subtlety depend on chemical reactions, on temperature-dependent phase changes (e.g., water changing to ice), and on life itself. For instance, the rate at which our landscapes are eroded is limited not by the capacity of running water to carry sediments, but also by the rate of rock "weathering": the chemical reactions with the air, water, and secretions of microscopic organisms that cause rock to become weak and susceptible to breaking and being washed away. Should we carelessly destroy the ozone layer and thereby do away with life on Earth, the atmosphere itself, which partly has been generated by life, might be altered. Weathering processes might thereby be modified, changing the familiar geological processes on our planet. This is speculation, but a planetary ecosystem is very intricate. How stable is the Earth's atmosphere, ocean, biology, and geology as a total system? We are still too firmly grounded in uniformitarianism to dare to find out.

Some inquisitive scientists have studied the stability of other planets, where we have fewer preconceptions and less to risk in discovering instability. Actually instabilities are common in nature, both the one-way kind (an avalanche) and the oscillatory kind (the cycle of ice ages). A more drastic form of climatic instability has been proposed for Mars by Cornell University astronomer Carl Sagan and his associates. Mars is now a cold, dry world with a tenuous atmosphere having a pressure that is only 2 percent of that at the top of Mount Everest. Yet Mars probably had huge amounts of water, other liquids, and gases when it formed. Sagan hypothesized that the polar ice caps on Mars might contain enough volatiles frozen out of the atmosphere (including water ice and dry ice) to constitute an Earth-like atmosphere if released. He calculated that any slight warming of Mars would begin a melting process that would eventually evaporate the ice caps and provide Mars with a thick atmosphere. Later the process might reverse. So Mars may have experienced one or more cycles of oscillation between its present state and an Earth-like state.

The effect of such climatological instability on the possibility for life on Mars is obvious, but the effect on Martian geology is equally significant. As described earlier, Mars shows evidence for a past episode of geological activity, including running rivers, which could be accounted for by the massive atmosphere that Sagan proposes. Other scientists doubt that sufficient volatiles are accessible to the atmosphere. But the correctness of Sagan's hypothesis is less important than the recognition it inspires—that analogous catastrophic processes might well have occurred on planets in times past, including our own. We certainly cannot assume past uniformity nor assume that other planets behave similarly to Earth.

The greatest catastrophes of all are those of colliding worlds. In one instant 4 billion years ago, more than half a billion years after the moon was created, an asteroid several dozen kilometers across struck the moon, forming the gigantic Mare Imbrium basin, which is bigger than Alaska. This moon-shattering explosion produced ridges and valleys a quarter of the way around the moon similar in size to the Appalachian ridges in Pennsylvania. Yet this event may have been only one of dozens of similar collisions and countless smaller ones that affected the inner planets during the so-called Great Cataclysm.

All catastrophes pale in comparison to the hypothesized Great Cataclysm. Whether it actually happened is not yet known, but eminent space scientists such as Gerald Wasserburg of the California Institute of Technology, George Wetherill of the Carnegie Institution, and Bruce Murray, now director of the Jet Propulsion Laboratory, have proposed that nearly all of the great craters on the terrestrial planets were formed during one relatively brief episode lasting less than 1 percent of solar system history! Such a fantastic bombardment-blizzard in the inner solar system would also have cratered the Earth and Venus and shattered the asteroids.

As astronomers now understand the collisional and orbital evolution of the solar system, this catastrophist's dream is not unexpected. For instance, collisions between minor planets, or asteroids, certainly happen from time to time in the asteroid belt between the orbits of Mars and Jupiter. Had one of the larger asteroids in an appropriate orbit been destroyed, the fragments might have been sprayed throughout the inner solar system. They could have created all large craters visible on Mars, Mercury, and the moon, as well as a similar number on the Earth (since eroded away) and on Venus.

George Wetherill has suggested another plausible scenario for the cataclysm. The orbit of a large asteroid-sized body originating beyond the orbit of the planet Uranus in the outer reaches of the solar system might have been gravitationally perturbed, just as comets are today, into a highly elongated orbit so that it repeatedly entered the inner solar system. During one orbit it is entirely possible that it may have barely missed hitting a large planet and would have been torn apart by wrenching tidal forces. (The chances of passing close enough to Earth to be disrupted by tides are ten times the chances of actually impacting.) The pieces of the disrupted body might then have cratered all the terrestrial planets.

While both suggestions for producing the Great Cataclysm involve a chance event in the distant past, there is nothing improbable about such close planetary approaches or asteroid collisions. Indeed a similar event could happen in the future. Fragments from the 300-kilometer-diameter asteroid Davida, should it ever suffer a catastrophic collision, might well produce another great cratering episode. And it could well get struck sometime in the next few billion years. Craters the size of Meteor Crater would be formed somewhere on Earth every few decades and a few impacts might trigger Earth-enveloping quakes. But we don't need to start worrying about such a celestial disaster; at last account Davida was still doing fine out in the asteroid belt, and even were it zapped today, it would be a few thousand years before any of its fragments ventured to within striking distance of Earth.

Having seen how easily a Great Cataclysm could have occurred, let us consider whether in fact it ever did occur. It is surely ironic that many of the arguments for such a catastrophe are grounded in—you guessed it—uniformitarianism. The cataclysm was first proposed by a group of geochronologists at the California Institute of Technology who call themselves the "Lunatic Asylum." Evidently the King of Hearts is Professor Gerald Wasserburg, who heads the National Academy of Sciences' Committee on Planetary Exploration, so the inmates' proposals must be taken seriously. Few lunar rocks date from before 4 billion years ago, and the cutoff is very sharp between plentiful, younger rocks and absent, older ones. Since many old moon rocks have traits indicating that they have been greatly smashed, the geochronologists ascribed the cutoff in rock ages to a "terminal lunar cataclysm" —an episode of saturation bombardment that melted or otherwise destroyed all preexisting rocks.

Furthermore, while ages of moon rocks from several different sites differ only slightly, the accumulated numbers of craters in these areas—as counted from photographs—differ greatly. Proponents of the cataclysm argue that the lava flows that formed these surfaces occurred just before the end of the bombardment, so that regions only slightly older received many more impacts than regions formed a little later. Other scientists counter that many of the craters compared are not impact craters at all, so that no cratering blizzard is implied. Instead, they say, there was a gradual decline in the cratering rate during the final stage in accretion of the moonlets that formed the moon itself.

While arguments continue about the enormity and brevity of the lunar cratering bombardment 4 billion years ago, the controversy has spread to whether or not there is evidence for a simultaneous bombardment of Mars and Mercury. For these planets we have as yet no rocks to date, so how do we know when their craters were formed? Mars is everywhere less densely cratered than the lunar uplands, despite its proximity to the ostensible source of cratering projectiles—the asteroids. Couldn't the craters we see have formed continuously through Martian history, without any early bombardment?

Not according to Larry Soderblom of the U.S. Geological Survey in Flagstaff, Arizona. Soderblom is a very bright young planetologist who received a classical geology education. In December 1973 he gave an authoritative presentation before the International Colloquium on Mars, held in Pasadena, California. He set out to "compare the lunar and Martian cratering histories by studying the rate of change of Martian impact flux implied in the geologic record provided by Mariner 9, assuming uniformitarianism in the geologic evolution of Mars." By assuming that landforming processes on Mars occur roughly uniformly throughout time, Soderblom argued that the crude division of the Martian surface into heavily cratered landscapes and lightly cratered ones implied a sharply declining cratering rate near the beginning of Martian history.

Soderblom rarely uses the word *uniformitarian* and is not a vociferous advocate of uniformitarianism. But like any good uniformitarian, he feels intuitively that the burden of proof that Martian geology has *not* been uniform rests with his critics. With one further uniformitarian jump in logic, the early cratering on Mars is ascribed to the same source as the early lunar cratering: if the lunar cratering was cataclysmic, it was on Mars as well.

What about Mercury? The Mariner 10 television team, led by Bruce Murray, adopted a working model for understanding Mercury's heavily and lightly cratered provinces that rests largely on analogy with the moon and Mars: ". . . we find it plausible to correlate the terminal bombardments on both the Moon and Mercury as resulting from a distinct episode that affected at least the Inner Solar System about 4 billion years ago. . . . But Mars also exhibits a heavily cratered surface. . . . The simplest and, to us, the most plausible explanation is that all three surfaces record the same episode of solar system bombardment."

Of course, the Great Cataclysm is plausible. But so is the previously prevailing idea that most craters on planets represent the tail end of the formation of the planets themselves. If they formed by the impact of smaller planetesimals that were in nearly circular orbits in the planet's own zone of the solar system rather than in elongated orbits crossing the paths of several planets, then the cratering need not have happened in the same manner on each of the three worlds. In fact, one might expect their cratering histories to have been quite different.

The creation of scientific hypotheses involves a complex interplay of the data with scientists' intelligence, special expertise, and assumptions. When syntheses involving different fields are required—say celestial mechanics and geology—then scientists depend upon mutual communication and faith in each other's expertise. Philosophical assumptions and faith in the judgment of others often become preeminent when questions to be resolved are complex, data are recent and only superficially examined, and implications of the answers are fundamental.

A case in point is the recent development of an approximate "planetary cratering chronology" from the hypothesized cataclysm to the present. In the absence of datable rocks from Mars and Mercury there is only one way to establish an "interplanetary correlation of geologic time" that connects the relative sequences of geology on these planets to absolute dates known only for the Earth and moon: to understand how the quantities and orbits of the asteroids, comets, and other impactors have evolved since the hypothesized cataclysm. If we were to know, for instance, that there have been many more asteroids crashing onto Mars than onto the moon, then the fact that Mars has similar numbers of craters would imply it is a younger, geologically active planet. Unfortunately, we don't know the cratering rates from planet to planet to within a factor of 10; that is, Mars may be struck by ten times as many bodies as Earth, or perhaps just the same number.

Still, Murray, Soderblom, and Wetherill have tried to fashion a rough chronology for Mercury and Mars in this way: first, Larry Soderblom compared the crater densities on the moon to those on Mars and wrote, "Because the oldest, postaccretional [flat], surfaces on Mars and the moon display about the same crater density, it now appears that the impact fluxes at Mars and the moon have been roughly the same over the last 4 billion years."

And if the impact rates have been the same, then any province on Mars cratered similarly to a province of known age on the moon has a similar age. This analogy seems simple and straightforward, yet it arbitrarily excludes the possibility of differing fluxes and differing chronologies.

Bruce Murray carried the uniformitarian analogy over to Mercury. He observed from Mariner 10 pictures that "the light cratering on the flooded plains of Mercury is similar to that on the maria [plains] of the moon." He drew an analogy like Soderblom's, supported by the fact that "similar flux histories for Mars and the moon were independently hypothesized by Soderblom." But Bruce Murray, who has been chief of more planetary-imagery spacecraft experiments than anyone else, realized he further needed to know the relative number of impacts on different planets. So he turned to an article by George Wetherill, a man who developed his reputation in geochemistry and geochronology but who has evolved into an expert in celestial dynamics. Murray noted that "relatively uniform impact flux histories throughout the inner solar system for the last 3 to 4 billion years were inferred recently by Wetherill, who concluded that the impacting objects probably originated in [highly elongated] orbits" that would cross the orbits of all the planets. Thus Murray concluded that a "straightforward interpretation" supports a similarity in Martian, lunar, and Mercurian impact fluxes.

Murray's conclusion is perhaps less secure than he seemed to realize. First, Wetherill studied the orbits of many different populations of bodies that could have cratered the planets, none of which has identical impact rates on each of the terrestrial planets. But Wetherill had heard that "recent observations of Mars and Mercury . . . have suggested to several workers [Soderblom and Murray] the hypothesis that . . . all of the terrestrial planets have had essentially the same bombardment history." Wetherill particularly wanted to emphasize that, among all the kinds of bodies he studied, "the only bodies which [produce] a near equality of flux on the moon and all the terrestrial planets are those derived from the vicinity of Uranus and Neptune." He went on to conclude that they were responsible for many of the craters, thus supporting the "independent" interpretations of Murray and Soderblom. Thus a plausible but uncertain hypothesis seems to be independently confirmed, when in fact the scientists are relying on each other's "proofs" of equality in planetary cratering. In the end, uniformitarianism reigns.

I would be amiss in discussing solar system catastrophism not to mention Immanuel Velikovsky. Inspired by professional interests in psychoanalysis, he performed a solitary, unparalleled synthesis of the implications of biblical and other ancient myths and legends for solar system history. Although based less on rigorous logic than on analogy and circumstantial evidence, his results were truly revolutionary. For instance, he proposed that Venus and Mars hurtled past the Earth in historical times, producing the devastating disasters recorded in legends from around the world. Velikovsky would have Venus launched from Jupiter as a monster comet shortly before it encroached on Earth, releasing manna from the heavens and causing the Earth to cease turning (among other feats), a phenomenon that would then account for the parting of the Red Sea for Moses and the Israelites.

His monograph *Worlds in Collision* was published in 1950, over objections of many astronomers who tried to censor his ideas, and was a best seller. Velikovsky has since been shunned by the scientific establishment, though more recently he gained a forum in the now defunct magazine *Pensée.* Although Velikovsky is an extreme catastrophist, is his work unscientific? The most frequent criticisms of him are that his assumptions and reasoning are unsound, or simply that he is wrong. But many scientists reason by analogy, unfounded assumptions, and circumstantial evidence, and yet they are not barred from publishing in established journals. Nor has being wrong been an obstacle to publication: most articles published in decades past seem wrong to us now. Truth is hard to come by.

Of course, Velikovsky probably is wrong. It is impossible to imagine that one man could single-handedly refute most twentieth-century science without slipping up somewhere. And his supporters' claims that many of his 1950 predictions have been verified by subsequent research are simply false. The reason establishment scientists reject Velikovsky is that they believe him to be *so* wrong that they aren't interested in any research bearing on his hypotheses. Of course, that is not a very open-minded position to be taken by those who are supposedly searching objectively for truth.

The blame lies equally with Velikovsky, however. If it is not required that scientific research be correct or even tightly reasoned, it does matter that it be relevant and useful to the scientific process. It must be related to previous work or demonstrate past errors. Research that rejects the fundamental assumptions or paradigms of modern science (as Velikovsky's seems to) must show that the new assumptions provide a better explanation for all facts, not just a few. Yet Velikovsky has refused to place his theories in a context that can broaden the perspectives of, or be useful to, other scientists. He has not adequately defended the seeming incompatibility of his work with the laws of physics. Should he reject those laws, it will take much more than his solitary efforts to devise an acceptable alternative paradigm. Against his "take it or leave it" attitude, it is understandable that scientists wish not to devote precious space in their journals to publishing articles that have no bearing on their own work.

This does not mean that Velikovsky's challenge should be dismissed lightly. Do we really know that it is physically impossible for Venus to be in its present orbit, apparently stable for millions to billions of years, if it grazed the Earth only a few thousand years ago? When pressed, most specialists in orbital theory admit that theorems of orbital stability in systems as complex as the solar system have yet to be rigorously proven. And many of our other facts about the solar system are also built on assumptions not quite so fundamental as Newton/Einstein mechanics. We all ought to be more aware of our assumptions and more open to questioning them should contradictions appear. Or, at the very least, while most scientists pursue research guided by currently accepted paradigms, we should be tolerant of the few that march to different drummers.

For the near future, at least, Earth and planetary science will fashion its own mixed blend of uniformitarianism and catastrophism, rooted in the physical and chemical laws we think are fundamental. If and when our emerging model for the solar system's origin and evolution is found wanting, it may be overthrown as previous scientific paradigms have been (e.g., the recent acceptance of continental drift). It seems extremely unlikely to me that Velikovskyism will have been bolstered to the degree that it then will provide a superior model for the solar system, but who can say for sure?

As the Earth orbits the sun, it constantly encounters swarms of meteoritic grains left over from comets and asteroids. This $3\frac{1}{2}$-minute exposure by Dennis Milon was taken at the height of this century's greatest meteor shower on November 17, 1966. The meteors appear to be radiating from a point in the head of the constellation Leo, vividly demonstrating the Earth's hurtling trip through space. The star images, including Regulus just below center, appear trailed because of the Earth's rotation. The two point meteors near the radiant are flying straight toward the camera. *Courtesy Dennis Milon*

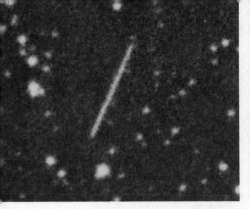

ABOVE LEFT: The streak passing through this field of stars is the tiny asteroid 1976AA. This 900-meter-diameter chunk of rock travels about the sun in slightly less than a year along an orbit very similar to the Earth's. It was discovered in January 1976 on this plate by E. Helin and E. Shoemaker. *Hale Observatories photograph, courtesy E. Shoemaker and E. Helin*

ABOVE RIGHT: Almost everywhere on Earth rainfall and running water shape the land. But here, in the Iqa Valley of Peru, virtually no rain falls and the landforms are shaped mainly by blowing winds and sand. These elongated ridges, several kilometers long, are called yardangs; their aerodynamic shapes resemble the bottom of a boat. While confined to the driest deserts on Earth, yardangs seem to be common on Mars. *Courtesy NASA*

BELOW: The crater Elegante lies in the Pinacate lava fields, south of the Arizona-Mexico border near the Gulf of California. Unlike the smaller cinder cones, or volcanic mountains, visible at the top, Elegante bears at least a superficial resemblance to meteor-impact craters. But it, and a number of other similar craters in the Pinacates, are known to be of volcanic origin, so it is easy to understand why debate raged so long about whether the lunar craters are of impact or volcanic origin. *Courtesy S. Larson*

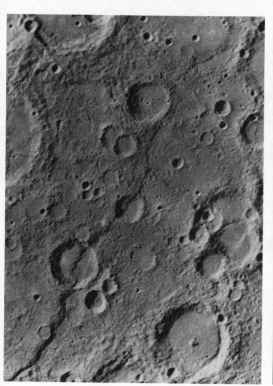

ABOVE LEFT: Discovery Scarp is the prominent feature that runs north to south on the right side of this Mariner 10 photograph of Mercury. It was named after the sailing ship used by Captain Cook on his 1776–80 explorations of the Pacific, Canada, Alaska, and Siberia. It is one of the numerous wrinkles on the surface of Mercury that provide evidence of the contractions of the planet as it cooled. *Courtesy NASA*

ABOVE RIGHT: The left half of this mosaic of Mariner 10 photographs is dominated by the huge Caloris Basin, which is over 1,300 kilometers in diameter and is the largest basin on the side of Mercury observed during the mission. The basin's floor is covered with ridges and canyons. The formation of the lightly cratered plains to the right may have been associated with the formation of the basin itself. *Courtesy NASA*

45

ABOVE: This close-up view of Mercury is of a surface that measures only 50 by 40 kilometers. It was taken when Mariner 10 was less then 6,000 kilometers from the planet. Most of the craters visible, some only hundreds of feet across, are subdued or "softened" in appearance. Perhaps they were originally fresh but have been eroded over the eons. Or maybe they were formed soft, either by the low-velocity impact of blocks ejected from nearby primary impact craters or by volcanic processes. Note that many of the craters are aligned in chains. *Courtesy Jet Propulsion Laboratory, NASA*

BELOW: This densely cratered region near Mercury's south pole was photographed during Mariner 10's second encounter with the innermost planet on September 21, 1974, when the spacecraft was 76,000 kilometers from Mercury's surface. A major lobate scarp curves down through the center of the picture. *Courtesy Jet Propulsion Laboratory, NASA*

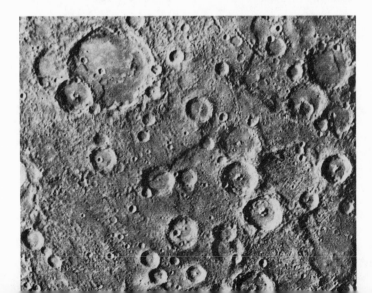

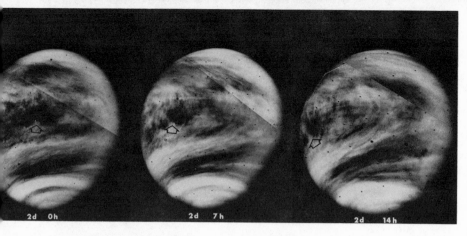

ABOVE: Photographs of Venus taken at 7-hour intervals 2 days after Mariner 10 flew past the planet en route to Mercury in 1974. The cloud patterns in Venus's upper atmosphere, distinguished in ultraviolet light, are rotating from right to left. The swirling cloud patterns in the stratosphere are especially prominent toward the poles. The arrow indicates a dark patch about 1,000 kilometers across. *Courtesy NASA*

BELOW: An approximate map of the cloud patterns on Venus has been constructed by piecing together strips from 33 Mariner 10 photographs taken during a 7-day period following encounter. The largest patterns, such as the dark, horizontal, Y-shaped feature that surrounds the planet, persist for periods much longer than the roughly 4-day period of rotation. Note that similar patterns are visible at 4-day intervals; the "map" goes nearly twice around the planet. *Courtesy M. J. S. Belton*

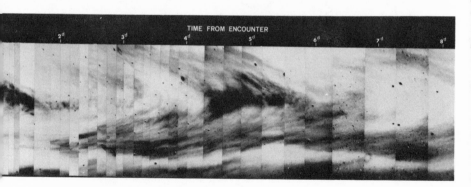

A Russian New Year's greeting from the surface of Venus. The cartoon figure sits on a part of the Venera 9 spacecraft, which landed on the hot (500 degrees C) surface of Venus in late 1975. Pebbles and rocks are scattered on the landscape toward the horizon, which is visible in the upper right-hand corner.

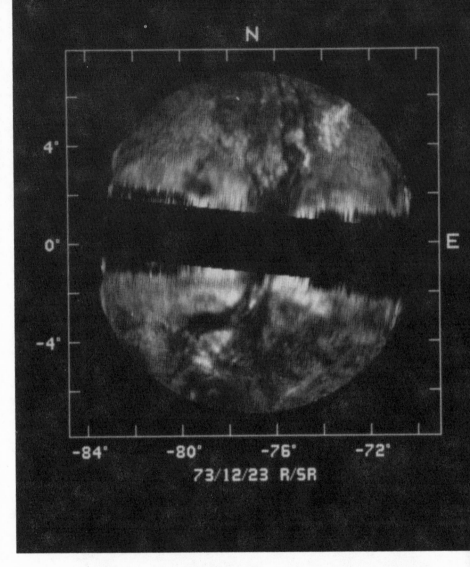

Radar map of circular patches on the surface of Venus. The Goldstone radar echoes penetrated the thick Venusian atmosphere, permitting geologists to map hills and valleys on the planet's surface. Each patch is about 1,500 kilometers across, with a band of poorer-quality data blacked out in the center. This view shows a giant, branching canyon running north to south, with a scalloped tributary canyon running off the top of the mapped area. *Courtesy R. M. Goldstein, H. C. Rumsey, and R. R. Green (JPL radar observations are supported by NASA)*

ABOVE: The Mariner 10 spacecraft. The cameras that took pictures of Mercury and Venus peer down from the top of the device. The "wings" are solar panels, each nearly 3 meters long. *Courtesy Jet Propulsion Laboratory, NASA*

BELOW: This oblique view of the crater Eratosthenes was snapped from the moon-orbiting Command Service Module of the Apollo 12 lunar-landing mission. The crater is surrounded by clusters of smaller, secondary craters formed by the impact of material ejected during the formation of Eratosthenes itself. *Courtesy NASA*

A chain of craters photographed from the orbiting Apollo 14 Command Module. *Courtesy NASA*

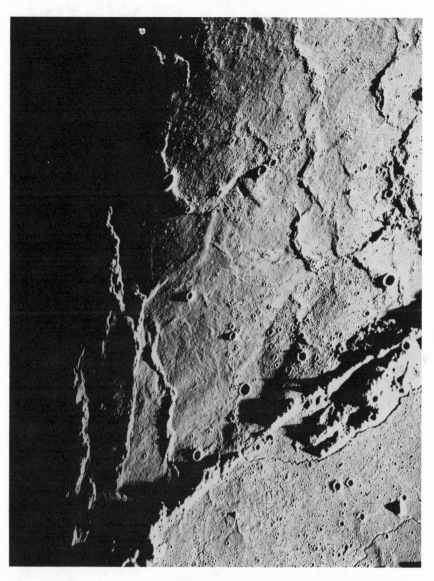

As the sunlight gently brushes the lunar surface near the day/night boundary, the rather smooth lunar surface takes on an exaggeratedly rough appearance. This far-side picture was taken from Apollo 15, which visited the Hadley Rille in summer 1971. *Courtesy NASA*

ABOVE: The giant, flat-floored crater Ptolemaeus, located near the center of the Earth-facing side of our moon, as photographed by Apollo 16. The sun's low angle of illumination enhances the profusion of tiny craters visible on the floor of this 150-kilometer-diameter basin. *Courtesy NASA*

BELOW: An unfamiliar view of the moon. This far-side photograph was taken toward the end of Apollo 16's visit to the moon. *Courtesy NASA*

4

Fragments from the Solar System's Birth

OUR CULTURE VALUES greatness. Things small but beautiful never make the famous *Guinness Book of World Records*. Every touchdown leads inexorably to the Super Bowl. We extol superstars among us, one of whom stares from *Time*'s cover each new year as Man (occasionally Woman) of the Year. No wonder we all know of Jupiter (the largest planet), Venus (the nearest planet), and Saturn (the only ringed planet). But who has heard of the planets Bamberga, Dembowska, Juno, or Nysa? They also orbit the sun but are diminutive in size—just four among the countless asteroids that range in diameter from the breadth of Texas down to the mere pebbles and celestial sand grains we call meteoroids.

The French aviator and poet Antoine de Saint-Exupéry challenged our preoccupation with matters of greatness in his fable *The Little Prince.* His young hero lived on a small asteroid and cared for a single, beautiful flower. The prince traveled to Earth and en route visited some minor planets, numbered 325, 326, 327,

328, 329, and 330. It was with prescience that in 1943 Saint-Exupéry portrayed each as a small, unique world, with a single inhabitant of characteristic virtues and vices. The Little Prince met a benevolent king, a conceited man, a rationalizing drunkard, a greedy capitalist, a devoted lamplighter, and finally a narrow-minded scientific specialist. Befitting the variety of their occupants, some of the tiny worlds were bigger, others smaller; some spun rapidly, others more slowly; some had mountains, others not; and they were of many different hues.

It has only been in the 1970s that astronomers too have begun to realize the variety of these little worlds and their significance for understanding our own. Although the discovery of Ceres, the largest asteroid, in the year 1801 was greeted with enthusiasm, the planet was disappointingly small for that which had been predicted to lie between Mars and Jupiter. As ever smaller asteroids were found, interest in them degenerated to a mere race among the less imaginative observers to discover more than anybody else. Saint-Exupéry satirized these endeavors by remarking that the discovery report of "asteroid B-612" before an international conference was by an astronomer in "Turkish costume," so the report was not believed. More than one astronomer has called asteroids "the vermin of the skies" because they meander through familiar star patterns leaving unappealing trailed images on long-exposure photographs of galaxies and other (to some) more glorious celestial wonders.

The growing respectability of asteroid research is largely the result of the efforts of Tom Gehrels. Now a professor at the University of Arizona, Gehrels pays attention to details and, like the Little Prince, loves to travel to out-of-the-way places. It was on such a trip to America as a youth from the Netherlands that he wandered up Mount Palomar and was introduced to the astronomers by the groundskeeper, who had found him and put him up for the night. Nowadays he and his family may journey to Borneo or the Sierra Madre, but his most contemplative and relaxing times are spent once again on Mount Palomar, where he periodically photographs the skies in search of faint and unusual asteroids.

Tom Gehrels works now mainly with an instrument on two Pioneer spacecraft that were launched to explore the giant planets Jupiter and Saturn. As Principal Investigator of this experiment, he has been responsible for the close-up color photographs of the Red Spot and swirling cloud belts on the giant planet. Yet his interest in Jupiter is but a passing fancy compared with his two decades of work on minor planets. Together with his mentor—and founder of modern planetary astronomy—Gerard Kuiper, Gehrels has patiently pursued since the mid-fifties a program of measuring the brightnesses and colors of asteroids. The modern realization that minor planets are a multitude of individual bodies, with different colors, different shapes, and different rates of spin, has grown from this program of photometry.

By recording the brightness of an asteroid with a photomulti-plier—a supersensitive electric eye—attached to a telescope, Gehrels can measure a "light curve." As an oblong body spins on its axis, it first seems brighter as it presents its broadest sunlit face to Earth, then fainter as we see it end-on, then brighter again, and so on. The more rapidly the brightness fluctuates, the faster the asteroid must be spinning in space. The Little Prince, who loved sunsets, could watch more than 10 each 24 hours if he were on Icarus (asteroid 1566), but barely more than 1 in 24 hours if he were on the larger, pinkish body called 532 Herculina. Especially great brightness fluctuations reveal an asteroid's shape to be unusually oblong or irregular. Gehrels has even devised a way to calculate the direction in which an asteroid's north pole is pointing by measuring the light curve during different years.

Gehrels was ahead of his time. As late as 1970, by which time he had assembled a small group of students and Tucson high school science teachers to assist his research, he remained the only astronomer in the world seriously studying the physical properties of asteroids. To be sure, he had gathered statistics on spin rates and colors of dozens of asteroids, ranging in size from the small mountains hurtling past Earth to the much larger, more enigmatic

"Trojan" asteroids orbiting the sun on each side of Jupiter in that planet's distant orbit. But what did all these data mean? What did they reveal about what asteroids were made of and how they were formed? Gehrels didn't really know, and at the time nobody else was very interested.

By 1970 a few graduate students at other universities had begun to take an interest in asteroids. The Nobel Prize–winning Swedish physicist Hannes Alfvén, who founded the esoteric subject of magnetohydrodynamics and was studying its possible ramifications for the origin of small bodies in the solar system, publicly rebuked the National Aeronautics and Space Administration for ignoring asteroids as potential targets of future spacecraft missions. Gehrels decided the time was ripe and, with Alfvén and several others, he organized a three-day international symposium on the physical studies of asteroids. Held in Tucson, Arizona, in March 1971, the conference was attended by over 140 scientists, who presented 70 talks on various aspects of asteroid science. In the preface to the 700-page book of symposium proceedings, Gehrels wrote, "We are now on the threshold of a new era of asteroid studies. . . . Physical studies . . . have not been popular, at least not among astronomers. The lack of appreciation is coming to an end with the presently growing realization that asteroids, comets, and meteoritic matter are basic building blocks of the original solar nebula. Their exploration gives data for the study of the origin and history of the solar system. [This international conference was organized] to promote new and increased exploration [of asteroids]. . . ."

The conference was an exciting one for participants, who discovered the breadth of their mutual, latent interest in these neglected members of the solar system. Since then, Tom Gehrels's hopes have proven prophetic: his conference ushered in a period of frenzied research activity. Asteroids, once the purview of a single researcher, now account for 10 to 20 percent of ground-based planetary astronomical research.

What caused the moon to melt early in its history? Why are there nickel-iron meteorites, hunks of solid metal fallen from the skies which were used in centuries past both as anvils and as objects of devotion? How were the precious pre–solar system remnants of stellar explosions, which some researchers claim to have found in meteorites, protected from the heat and collisions that melted, vaporized, or ground down to nothingness so much material once part of the early solar system? How did the vast whirlpool of dust and gas that surrounded the sun at its birth manage to agglomerate and coalesce into just a few dozen large planets and satellites? Only in the few years since Gehrels's symposium have astronomers fully appreciated that precious clues concerning such diverse questions reside in the asteroids—those neglected, rocky fragments beyond the orbit of Mars.

Consider the sizes of the minor planets. The largest is Ceres, with a diameter of about 960 kilometers. Two others exceed 500 kilometers. But 30 or 40 exceed 200 kilometers, the length of Massachusetts excluding Cape Cod, and roughly 3,000 exceed 20 kilometers (the size of Lake Tahoe). Literally trillions of uncharted boulders the size of a basketball or larger exist. For each asteroid there are 10 others one-third its size. It is difficult to visualize the ever-increasing numbers of smaller bodies. After all, most familiar things in our world are of similar sizes. Few apples in a basket, or pigeons in a park, differ in size by as much as a factor of 2. Yet one familiar process does produce such an asteroidlike size distribution. Parents of youngsters know all too well what happens when a window or priceless vase is smashed. Smash a brick with a sledgehammer and count the broken fragments. The more you repeatedly grind something to pieces, the more nearly the fragments have an asteroidal size distribution.

So it is clear that asteroids are fragments. Indeed it had been supposed since the first ones were discovered that they might be fragments of an exploded planet. Recently the Canadian expert on celestial mechanics Michael Ovenden suggested that the asteroids are the remains of a giant Saturn-sized planet he calls

Aztec, which mysteriously self-destructed about the time of the demise of dinosaurs and emergence of mammals on Earth, 10 or 20 million years ago. It boggles the mind to imagine what kind of catastrophe could blast apart such a huge planet, so tightly bound together by its own gravity. But there are other incredible aspects to Ovenden's scenario. Since the total of all asteroids today makes up only 1/100,000 of the proposed mass of Aztec, 99.999 percent of it must have been flung about the solar system and eventually ejected into interstellar space. Yet lunar lava flows, which have been passively recording impacts for more than $3\frac{1}{2}$ billion years, have been cratered only infrequently, and collected lunar soils show no chemical trace of Aztec. How could the moon have remained unscathed? Why do the remaining asteroids orbit the sun in the same direction in a relatively orderly ring, in roughly circular orbits, rather than scatter about in highly inclined and elliptical paths intersecting at the site of the supposed explosion?

Actually the fragmental nature of asteroids is much more simply understood as resulting from a kind of celestial demolition derby. One can calculate, simply by knowing how many asteroids there are, how fast they are moving, and what it takes to smash them to pieces, that they must be grinding themselves down to dust. Imagine you were to enter an arena to watch a dozen driverless jalopies careening about. Wouldn't you be surprised if the autos were all functioning and undented? Not if the event had started only seconds before you arrived. But after half an hour of this madness, most vehicles would certainly be out for repairs. Similarly, there are so many asteroids flying about in space at relative speeds of 5 kilometers per second that, if they are made of ordinary rock, nearly all of them must be fragments from the catastrophic destruction of preexisting bodies that have collided over the eons. Even if there were a thousand times as many asteroids when the solar system formed as there are today, they would have smashed each other to bits so rapidly because of the crowding that there still would be only as many asteroid fragments as we see today.

There's an important *if* in that chain of logic, however: asteroids are fragments *if* they are made of ordinary rock. (Were they as strong as steel, few collisions would shatter them.) Are they rocky? In the last few years we have learned the answer to that question: most of them are. We realize that we even have pieces of some of these asteroidal rocks right here on Earth—in our museum collections of meteorites, the stones that have fallen from the skies!

We know about asteroid compositions from research initiated by Tom McCord in 1969 and continued by both him and myself. A onetime truck driver and serviceman, McCord is a dynamic, self-made man who made the most of his chance—through the G.I. Bill of Rights—to go to college. He soon earned his doctorate from the California Institute of Technology (Caltech). Within a few short years, as Professor of Planetary Physics at the Massachusetts Institute of Technology (M.I.T.), he had formed a small empire called the Planetary Astronomy Laboratory (MITPAL), a half-million-dollar annual operation. (For obscure bureaucratic reasons the acronym MITPAL evolved to MITCRS, then to MITRSL. More recently the lab has moved to Honolulu and is no longer affiliated with M.I.T.) MITPAL's chief products are the couple of hundred articles published in the professional magazines describing progress on Tom McCord's major project: to determine the surface composition of virtually everything in the solar system by the technique of "narrow-band spectrophotometry." Such a major undertaking required the operational talents of a man like McCord, who happened on the scene at just the right time.

We see planets and asteroids by the reflected sunlight that illuminates them, but the sun's rays are modified while interacting with planetary surfaces. On a microscopic scale, the rays bounce around among the mineral grains on the rock surfaces and are transmitted through some of them before traveling back to our eye. Some rocks absorb most of the blue, violet, and ultraviolet rays and transmit easily only the orange and red rays; thus they appear reddish. Others strongly absorb the rays toward both the violet and red ends of the spectrum, yielding a greenish hue. Anyone who has hiked into the Grand Canyon, or just strolled along a rocky stream, has noticed the variety of colors of rocks on Earth. Such colors ultimately result from the atomic structure of the mineral crystals of which rocks are made. The outer electrons, particularly of iron, cobalt, and similar atoms, are unusually sensitive to light of particular wavelengths in the red and invisible infrared; the exact hue of absorbed light depends on the crystal structure of the particular mineral. McCord can infer the minerals that compose rocks on remote planets just by measuring the wavelengths of the more strongly absorbed reflected sunlight. Actually, to measure these spectra is not an easy task: MITPAL engineers must build instruments, called spectrophotometers, that are much more sensitive to fine gradations in colors than is the human eye. These instruments have been used on many of the great telescopes atop mountains in Hawaii, South America, and the Southwest, including the 200-inch Hale telescope on Mount Palomar.

The power of this research technique was proven when McCord demonstrated, months before astronauts returned the first moon rocks to Earth, that the moon's surface was made, at least in part, of the very same minerals that compose rocks on Earth! Of course the major minerals in lunar rocks occur in different proportions than in terrestrial rocks, revealing clues about the nature of the moon (see Chapter 8).

Spectrophotometry has been applied particularly successfully to the asteroids. Some contain minerals common in certain terrestrial rocks, but others are quite exotic. Consider asteroid 324 Bamberga. Perhaps the Little Prince began his travels with Number 325 because he couldn't even see Bamberga, which is dim and seems unexceptional. But in 1970, when then-Caltech graduate student Dennis Matson measured the heat being radiated toward Earth by Bamberga, he found it was one of the brightest sources of thermal radiation in the sky, which meant that it was a warm and unexpectedly large body. For Bamberga to be large and warm but dim to the eye means that it must be exceptionally black. In fact, it absorbs more than 98 percent of the sun's rays of all colors, from ultraviolet straight through infrared. Bamberga is blacker than coal dust!

What can a 230-kilometer-diameter planet be made of that is that black? Clearly it must contain chemical elements that cosmochemists have shown are reasonably abundant in the universe. The only common element that frequently forms black compounds is carbon, which is fourth in cosmic abundance, after hydrogen, helium, and oxygen. Yet even on Earth black carbon compounds are quite rare and usually are associated with decomposed life; from them we make "lead" pencils, black ink, carbon paper, and so on. But there is one kind of black, carbonaceous rock that falls from the sky: the carbonaceous chondritic meteorites. These meteorites, while rarely found because of their fragility, are believed to be the most common kind of rocks in space. McCord's MIT-PAL colleagues Torrence Johnson and Michael Gaffey measured the light reflected from powdered carbonaceous meteorites and found that it closely matches Bamberga's spectrum. So it seems likely that Bamberga and the many other similarly black asteroids are giant carbonaceous chondrites and that the carbonaceous meteorites in our museums are chips from some of these asteroids —extraterrestrial samples far more unearthly than any moon rocks!

Rocks in the Earth's crust are rich in silicon and aluminum. Spectrophotometry reveals, however, that such rocks are uncommon on asteroids, whereas many iron- and magnesium-rich minerals occur on the noncarbonaceous asteroids similar to those minerals that compose noncarbonaceous meteorites. A few asteroids contain a high percentage of the silicate minerals pyroxene and olivine, just like the ordinary chondritic meteorites. A couple of highly reflective asteroids seem to be made of whitish rocks like the unusual aubrite meteorites. Vesta, the third largest asteroid, is made of yet a different kind of rock that resembles the volcanic rocks that spread across the surfaces of the moon and Earth; in this respect Vesta is unique among asteroids, but as I will describe shortly, there may have been many Vesta-like bodies in earlier epochs.

The second most common type of asteroid—carbonaceous ones turn out to be most common—are moderately reflective and have a reddish tinge. Some of them contain rocky minerals, but MIT-PALers discovered that a major component of these asteroids is a pure nickel-iron metal alloy, unlike anything found naturally on Earth (except of course in the iron meteorites). If McCord is right, there are hundreds of millions of billions of tons of nickel-iron alloy in the asteroid belt. The economic potential of this storehouse of metal, in the event mankind conquers and industrializes space, is staggering. But the implications of McCord's interpretation for the early history of the solar system are also fascinating and important.

I discussed earlier the inevitable fragmentation of the asteroids and the assumption that they have strengths similar to ordinary rock. It turns out, indeed, that most asteroids are rocky, which confirms why asteroidal sizes are of a fragmental nature. But any asteroids of iron-nickel composition are stronger and much less easily smashed, so in the 4½ billion years of solar system history we should expect few of them to be fragmented. Hence the reddish, metallic asteroids that McCord and I have measured should not exhibit a fragmental size distribution. And indeed they don't.

One afternoon in August 1974 I graphed the sizes of the metallic asteroids separately from the rocky ones and I noticed a curious lack of smaller metallic asteroids. The absence was all the more striking because such metallic asteroids are more reflective than the black, carbonaceous ones and are therefore less easily overlooked. Within an hour or so it dawned on me that the nonfragmental character and implied strength of these bodies confirmed McCord's interpretation of a metallic composition.

What could preferentially form solid metallic bodies, roughly 100 to 200 kilometers in diameter, in the first place? Some years earlier, meteoritical researchers had already determined that iron-nickel meteorites must have been buried in parent bodies several hundred kilometers in diameter. Imagine there were once parent bodies composed of chondritic meteorites, which are believed to be the materials that originally condensed from the cloud of gas from which the solar system formed. If, somehow, such bodies were heated and melted, the heavy iron and nickel in them would have sunk to their centers, forming molten nickel-iron cores. Geochemists have calculated that the lightest materials would be volcanic rocks, which would float to the surface, just as on Vesta, the lava-covered asteroid mentioned before. After cooling for 100 million years, Vesta and other similar melted bodies would have solid metallic cores 100 to 200 kilometers in diameter, just the size of the metallic asteroids.

But what stripped away the volcanic crusts and rocky mantles from all of the melted asteroids besides Vesta so that we now see the metallic cores directly? I am reminded of winter days when I was growing up in Buffalo, New York, quite the opposite of the broiling August day in Arizona when my mind was feverishly piecing together this puzzle. Two brothers, about my age, who lived on the block had tendencies toward delinquency. They weren't satisfied, as the rest of us were, to throw fluffy snowballs at passing cars. They froze their snowballs into iceballs. Occasionally they made snowballs with rocks in the centers. It is easy for me to remember the rocky snowballs denting the cars and the irate drivers chasing the boys through the neighborhood. But for our present purposes, consider the fate of the snow and the central rocks. The snow, of course, splattered everywhere, but the rocks remained intact. Similarly, over eons in the celestial demolition derby, the odds have become very large against the survival of the rocky outer layers of the once-melted Vesta-like asteroids, but most of the 100- to 200-kilometer-diameter cores remained whole. Of course, the few that have suffered an unusually violent collision have given rise to some smaller metallic fragments, including the meteorites that fall on Earth. As for the original melted bodies, only Vesta has been so lucky that even its crust survived relatively unscathed.

Inevitably, data amassed since 1974 are not all in accord with this deceptively simple picture of asteroid collisional evolutions. Still, in just a few years, astronomers have learned much about what the asteroids are like: how big they are, what they are made of, and so on. Indeed we have learned that most meteorites almost certainly are derived from the asteroids—a venerable hypothesis that nevertheless was doubted by most theoreticians only a decade ago. We also have learned how the asteroids behave today, in particular the occasional collisions that, multiplied over eons, have pulverized them to a swarm of debris. We even have ideas about what the asteroids were like several billion years ago. But this knowledge only sets the stage for addressing more fundamental questions: why is there an asteroid belt, rather than a planet, between Mars and Jupiter? Why were some asteroids heated to the melting point while others seem to be composed of unmodified material from the solar system's birth? At last we can make some informed speculations, but final answers await research in years to come.

Let me portray a scenario for the origin of the asteroids and the larger planets. We are confident of parts of this picture, but many mysteries remain. The sun was born from a giant cloud of gas and dust left over from some exploding stars in the Milky Way galaxy. Called the solar nebula, the cloud contracted and the internal gas pressure increased; hence the temperature rose. By the time a proto-sun appeared, the swirling nebula had become a revolving disk of gas with temperatures of a couple of thousand degrees in the zone of the yet unborn inner planets. As it cooled, the gas began to condense into grains of minerals that can exist at the highest temperatures. Farther from the sun, the nebula cooled so that a wider variety of minerals condensed where the Earth and Mars were soon to form. Evidently carbonaceous chondritic matter, which completely condenses only at temperatures below a few hundred degrees, formed in the asteroidal zone. Farther outward it was still colder so that ices formed, leaving only the lightest and most volatile substances in gaseous form.

The grains grew from the condensing gas and, just as in a snowstorm, they began to settle and fall to the central plane of the nebular disk. Remaining gases were dispersed by the youthful sun, while throughout the central disk the mutual interactions of the crowded grains caused them to gather together into small spheres called planetesimals. These bodies in turn amassed into larger ones, perhaps 10 kilometers (6 miles) in diameter. Planetesimals near the sun, where Mercury was to form, consisted of only a tiny fraction of the original nebular gases—materials, including iron, that condense at the hottest temperatures. But farther out, where even the remaining gas had not been blown away by the distant sun, space was crowded with icy planetesimals that bumped into each other and grew in size. One of them, a proto-Jupiter, grew to the size of Ceres; its gravitational field began to attract gases and other planetesimals. In a cosmologically brief time the mighty planet Jupiter had formed.

Meanwhile, other planets slowly accreted from swarms of planetesimals. For a planet to grow, it was necessary for the planetesimals to move with respect to each other; otherwise they would never bump together. They could not be speeding too fast or they would fragment rather than grow together. So the asteroids, which eventually grew to hundreds of kilometers in size, must have originally been moving in similar, circular orbits, inclined only slightly to the flat central plane of the planetesimal system. Yet today they flash by each other at speeds of many kilometers per second and smash each other to bits when they collide. What stirred up their once orderly orbits, and was this mysterious occurrence the reason why the asteroids never formed a planet? Nobody knows yet, although we are beginning to narrow down the admissible hypotheses.

Many scientists think Jupiter must somehow have been responsible for the failure of the asteroids to accrete, as well as for the small size of the next-closest planet, Mars. Jupiter is so massive that its gravitational field can be felt at great distances. At certain distances from Jupiter, orbiting bodies pass by just often enough to receive periodic gravitational tugs from Jupiter that build up into large changes in their paths. Indeed, the regions in the asteroid belt that would have been inhabited by bodies orbiting with periods commensurate with Jupiter's mighty pulls are altogether absent. But in most parts of the asteroid belt, Jupiter's tugs occur at the wrong times and are not amplified into large excursions, so Jupiter's effects are calculated to be minimal.

Other theorists have suggested that the asteroidal planet was stillborn because of a cosmic shooting gallery created by Jupiter. Many icy planetesimals, left over in the outer solar system after Jupiter had formed, chanced to come near the giant planet; they were flung off in divergent paths, just as Jupiter tossed Pioneer 10 into interstellar space after its reconnaissance flight several years ago. Such speeding planetesimal-bullets entering the inner solar system might have been especially effective in smashing some of the nearby growing asteroids and Mars-zone planetesimals, thereby inhibiting growth.

Still, the scattering of planetesimals by Jupiter fails to account for the asteroids' divergent orbits. Imagine being given an arsenal of guns and being told to move a large glass sphere down the length of a football field by firing at it. If you fired at it with a BB-gun, it would move hardly at all; struck by a rifle bullet, the sphere would certainly be smashed to bits long before you got it to the 50-yard line. Similarly, asteroids are fragile and cannot be shifted intact by collisions. The stirring of the asteroids' orbits must have resulted from gravitational forces operating over tens of millions of years. Whether these were the mutual gravitational pulls on each other, or perturbations caused by Jupiter or other planets, we do not yet know. Whatever difficulties the asteroids had in forming a planet when the other planets were forming, the later, enigmatic disruption of their orderly orbits eliminated the possibility of any asteroidal planet.

We are fortunate that Aztec never formed, for we are thus left with slightly battered samples of planetesimals like those that originally formed the Earth, Mars, and Jupiter. All other evidence of this critical stage of planetary evolution has been lost forever as planetesimals were buried deep in the growing planets, melted, and transformed beyond recognition by the great pressures, geological forces, and chemical reactions endemic to large planets. Astronauts journeyed to the moon in hope of bringing back rocks from the earliest years of the solar system. But even our relatively small moon turned out to have been largely melted. Whatever early lunar rocks might have escaped melting are buried deep beneath kilometers of rubble on the lunar surface, since most of the rocks excavated by cratering impacts fall back onto the moon, rather than escape into interplanetary space as they would from a small, low-gravity asteroid. So the asteroids are the true Rosetta Stones of the solar system, and many of them probably contain rocks formed at the very beginning, when the solar nebula was beginning to condense. Most meteorites, the asteroidal fragments, date from roughly 4.6 billion years ago, making them more than 600 million years older than lunar rocks.

Most asteroids have been unmodified by thermal or chemical processes since the early condensation and accretion, but there are important exceptions. Remember that some dozens of asteroids like Vesta were heated; so much, in fact, that they melted and the minerals segregated into metallic cores surrounded by layers of lighter rocks. Why did they melt? And if some melted, why didn't the others? The answers may have wider implications for all planets. Asteroids are not easily melted since they are too small to generate high pressures in their interiors and they radiate away any radioactive heat almost as fast as it is produced. So if something melted asteroids, it probably would have had an even more important effect on heating the young moon, Earth, and other planets.

The sun, even if it were once much brighter than today, can only warm the uppermost layers of an asteroid. The radioactive decay of uranium, thorium, and potassium produces heat which, if trapped deep inside a planet, can eventually heat it to melting. This is the reason for the great internal warmth of the Earth today, but just as a thin-walled, poorly insulated house quickly loses heat during a cold night, tiny asteroids cannot be heated much by such decay of long-lived radioactive elements.

Some researchers hunting for the ancient furnace have turned to giant electrical and magnetic fields that may once have been carried past the asteroids by an interplanetary windstorm originating on the sun. Even today there is a stream of subatomic particles streaming away from the sun, but astrophysicists think this solar wind may have been much greater in our star's early history. In that case, calculations show that asteroid-sized bodies could be heated by the large currents induced by the passing magnetic fields, in a manner analogous to the way the elements on an electric stove are heated by flowing currents. A novel trait of solar-wind heating is that calculations show that asteroids of particular sizes and compositions may be heated but others may be affected little.

Another promising source of early planetary heating is the decay of radioactive elements having very short half-lives and which, therefore, are now extinct. For instance, one extinct form (isotope) of aluminum that contains 13 neutrons instead of the usual 14 tends to change spontaneously into a stable form of magnesium on a time scale of about a million years. Heat is a by-product of the decay. The unstable form of aluminum may well have been created by the stupendous stellar explosion, or supernova, that presumably created the cloud from which the solar system formed. If the solar system formed quickly enough—within a few million years of the supernova—the rapid radioactive decay of a sufficient quantity of the unstable aluminum might have provided a burst of heat for the newly formed planets in which it resided. Asteroids are large enough to retain heat for a few million years, if not for billions of years, so they also would have warmed and perhaps melted as a result. We can check to see if that happened by looking for unusual concentrations of the stable magnesium decay product in meteorites and lunar rocks.

Research a few years ago seemed to rule out such an early heat pulse, but more recently several research groups have reported anomalies in the proportions of magnesium isotopes in some meteorites that just may be the long-sought clue to the early thermal history of the planets. It will then be all the more important to

understand why some asteroids were heated to melting temperatures of over 1,000 degrees Kelvin while others never exceeded
their condensation temperatures of a few hundred degrees. Hang-
ing in the balance are some fundamental questions: how long did
it take the solar system, and the various planets in it, to form from
the debris of a stellar explosion? And how during the first 10
percent of the solar system's history did asteroidlike planetesimals
get together to form the planets—large worlds with the hetero-
geneous environments necessary for the evolution of life?

5

A World Revealed

The opportunity to add a whole new planet to our base of
knowledge about the terrestrial planets has been of
extraordinary importance, the intellectual implications
of which will continue to develop over succeeding years.

—BRUCE C. MURRAY, 1975

IT IS OFTEN REPEATED that Nicolaus Copernicus never saw the
planet Mercury. The tale is apocryphal, but of the five "wandering
stars" known to the ancients the winged messenger Mercury is
certainly the most elusive—a faint morning or evening star playing
tag with the sun and never visible in a completely dark sky. Now
geologists study Mercurian cliffs, ridges, and craters on aerial
photographs transmitted from a spacecraft that itself played tag
with Mercury. They are deciphering the history of this superficially moonlike world for which a decade ago we lacked even a
crude map.

The spasmodic history of our learning about Mercury exemplifies the scientific method, its mistakes and its accomplishments,
and especially the power of modern technology. Answers to some
perplexing planetary riddles are being uncovered by diligent analysis of reconnaissance data beamed back to Earth by Mariner 10.
Yet some of the most fundamental facts about the planet were

uncovered years before any spacecraft went near it, by Earth-based radar techniques that literally reached out and touched Mercury.

As schoolchildren, we all learned a few salient facts about the smallest planet, the one nearest the sun. First, tidal forces pulled one face of Mercury always sunward, so that its day exactly equaled its year of 88 Earth days. Naturally, the sunlit side was very hot and the dark side very cold. We learned, in fact, that Mercury was simultaneously the hottest and the coldest place in the solar system—a planetary baked Alaska. We may have heard of a tenuous atmosphere on this inhospitable planet and seen maps showing man-in-the-moon-like dark patches on Mercury's sunward hemisphere. These were the facts, undisputed well into the 1960s. The French astronomer Audouin Dollfus went so far as to assert that Mercury's rotation period was known to be 87.969 days to a precision of 1 part in 10,000.

But scientific facts, like so many others, are evanescent. To be sure, Mercury is nearest the sun, but although it is tinier than some moons of the outer planets, the distant planet Pluto is probably still smaller. And everything else we were taught about Mercury was sheer fiction. It is searingly hot at high noon on Mercury's equator, but not so hellish as practically everywhere on the surface of Venus. Nor does it ever get so cold on Mercury as in the outer solar system. The maps of Mercury's sunlit side were illusionary since there is no permanent sunlit side; Mercury does not keep the same face to the sun, but rather rotates once every 59 days or so. Nor is there even a tenuous atmosphere.

What went awry? It is a story of human fallibility, for Nature simply fooled us. It all started nearly a century ago with an Italian astronomer, G. V. Schiaparelli, at the Royal Observatory in Milan. He is famous—or infamous—for discovering the nonexistent canals on Mars. From that developed the conception of the Red Planet as the harborer of life, indeed of civilization. Should Mars turn out to have always been devoid of intelligent beings, just as today it is surely devoid of encircling canals, the twentieth-century view of Mars inspired by Schiaparelli's modest observations will nevertheless endure in human literature and history. But memory of his article "Sulla rotazione di Mercurio" ("On the rotation of Mercury"), published in 1890, will fade much sooner. And it is just as well, for if his observations of Mercury were conservative and his reasoning generally sound, they were also insufficiently perfect.

After tentatively observing the small rosy planet in 1881, Schiaparelli decided to make a regular study of it. He was the quintessential astronomer, peering at a twinkling planet through a long telescope and recording his observations in a diary, translated and paraphrased here: "I have observed Mercury in the telescope several hundred times and on more than 150 days it has been possible to see some spots on it, or at least something worth noticing. I have made about 150 drawings of it, which however are of uneven quality. But all more or less have contributed to the results of the present study."

Schiaparelli's logic in determining Mercury's rotation period from the positions of the spots had three elements which he conveniently numbered I, II, and III: "I. Observing Mercury on two consecutive days at the same hour . . . one sees the same spots . . . occupying approximately the same places on the apparent disk. . . . Of all the facts concerning the rotation of Mercury, this one is most manifest and the oldest known."

The German observer Johann Schroeter had made the same observation early in the nineteenth century. He had opted for the simple interpretation that Mercury rotates in 24 hours, just like the Earth. Schiaparelli pointed out two equally plausible alternatives. First, Mercury could make two or more complete rotations in 24 hours and present the same configuration of spots at precise 24-hour intervals. Or, it could rotate so slowly that the spots would not shift noticeably from day to day. Schiaparelli settled this question: "II. But observing the planet several times in the course of the same day at intervals of several hours, one still finds that its appearance is not changed. And the same is true when one repeats the observations on two consecutive days but at notably different hours, so that the interval is significantly greater or less than 24 hours. This fact is no less obvious than the preceding and is in open contradiction with the rotation of Schroeter. . . . Mercury rotates in neither a day nor a fraction of a day, but rather very slowly."

But with exactly what period? Schiaparelli's third and clinching observation was that even from year to year, the spots seemed to be reasonably fixed with respect to the boundary that divides Mercury's sunlit hemisphere from its night side. Only one conclusion seemed plausible; as Schiaparelli wrote, "The ensemble of these facts . . . shows that Mercury turns around the sun nearly in the same fashion as the moon around the Earth and Iapetus around Saturn, generally presenting to the sun (but with some oscillations) always the same hemisphere of its surface."

From the days of Kepler and Galileo, analogy has played a major role in planetary science. It was natural for Schroeter to have supposed Mercury's period to be 24 hours, like the Earth's, and just as natural for Schiaparelli to see a lunar analog once he had disproven Schroeter's result. Moreover, the reason for the moon's behavior had been proven mathematically just a decade earlier by George Darwin, son of the evolutionist; the theory seemed applicable to Mercury as well.

Schiaparelli's penchant for thinking of Mercury in Earth's terms made him overlook hints in his own diary that otherwise might have seeded doubts. First were his oscillations in positions of Mercury's spots; these he managed to ascribe to a well-known lunar phenomenon that would be greatly magnified for Mercury: libration. Kepler had shown that a planet in an elongated orbit, such as Mercury's, moves around the sun at an uneven rate. Yet a planet's spin about its axis is uniform, so the rotation gets ahead of, or lags behind, the day-night boundary, depending upon whether the planet is moving slower or faster in its orbit. A more obvious hint was that Schiaparelli's spots were "sometimes more visible and sometimes less." On occasion spots disappeared altogether, only to reappear a week later. Schiaparelli had a ready

Earth analogy for explaining these mutations: "It does not seem too rash to suppose that [these effects are due to] more or less opaque condensations produced in the atmosphere of Mercury that, from afar, appear analogous to the appearance that the Earth's atmosphere must present from a similar distance."

Later, during the 1920s, the great observer E. M. Antoniadi would likewise rely on veils and clouds to explain discrepancies between Mercury's appearance and his expectations based on the 88-day rotation period. Neither Schiaparelli nor Antoniadi considered that perhaps the spots, rather than being veiled, simply weren't there, having rotated around the planet. Still, the inconsistencies were infrequent and Schiaparelli can be forgiven for failing to attend to them. Usually Mercury does appear to behave just the way Schiaparelli reported—at least when astronomers are looking at it. And there's the rub! For Mercury cannot be kept under constant surveillance. It can be seen well only relatively briefly, during the six times a year it swings away from the sun. Three times it swings east of the sun and can be seen in our evening sky, and three times west into the predawn sky. From northern latitudes only two of these—an evening apparition in the spring and a morning one in autumn—provide superior views; at other times the image of Mercury is distorted by the turbulent, hazy air near the horizon. The interval between two such opportunities to observe Mercury's morning side (during evening apparitions) is just about 348 days, the same interval as between successive times when its evening side is favorably displayed.

Anyone who has seen wagon wheels appear to turn backward in a motion picture is aware of the illusions possible when a rotating object is viewed intermittently. This stroboscopic effect is as relevant to Mercury viewed every 348 days as to wheel spokes viewed each $\frac{1}{4}$ second. The stroboscopic illusion works only if the rotating object has a period very nearly two, three, or some other integral number of times the viewing interval. A filmed stage-coach's wheels misbehave only when the coach moves just so fast; most often the wheel spokes are a blur. It just so happens that exactly six of Mercury's 58.65-day rotation periods take 352 days, just 4 days longer than the 348-day viewing interval. Since exactly four of the supposed 88-day periods also take 352 days, it is easy to imagine that Schiaparelli's observations would be equally consistent with the true period and the lunar analog period he adopted.

Schiaparelli also overlooked other periods that are consistent with his observation III. They are other whole-number fractions of the viewing interval ($\frac{348}{3} = 116$; $\frac{348}{5} = 70$; $\frac{348}{7} = 50$; etc.). The 50-day period was a particularly serious omission, for with such a period Mercury would not only present the same face to Earth each favorable apparition but at every apparition, just as for the 88-day period Schiaparelli chose.

One further numerical coincidence helped fool Mercury observers. The 348-day interval between every third apparition is roughly equal to one Earth year. But it is not exactly 365 days, so the favorable spring apparitions occur about two weeks earlier each succeeding year; after 6 years or so they are no longer favorable spring apparitions but become mediocre winter apparitions, and eventually unfavorable autumn ones. Schiaparelli quit studying Mercury after 7 years, before the shifting apparitions brought new spots into view. And E. M. Antoniadi, who wrote the definitive book on Mercury, happened to observe only during the years 1924 to 1929, so there is little wonder that he confirmed Schiaparelli's result.

There is no reason at all for Mercury's true rotation period to be linked in any way to periods of the planet's visibility from Earth, so astronomers accepted the imperfect logic of Schiaparelli and Antoniadi. There was the lingering question of how this tiny, hot world could keep an atmosphere, required for the existence of reported veils, from evaporating into space. But Audouin Dollfus came to the rescue in the 1950s when he claimed—fallaciously, it turns out—that some properties of polarized light reflected from Mercury indeed proved the existence of a thin atmosphere.

We entered the Space Age secure in our portrait of the innermost planet. Complacency gave way to amorphous uneasiness in the early 1960s when improved radio-astronomical technology permitted detection of radio emission from Mercury. A warm body radiates at all wavelengths, mostly in the "thermal infrared," but also at very long wavelengths called "radio." Mercury was a thin crescent and presented mainly its dark side to the radio telescope. But the surprisingly strong radiation revealed that Mercury's night side was warmer than the perpetually frigid temperatures expected. Mysteriously, the warmth from the sunlit side seemed to be leaking around, or through, the planet. Nobody considered that the supposed eternally dark side might have been basking in the broiling heat of the sun only a couple of months earlier.

The clincher came in 1965 from radar technology. A natural amphitheater in the Puerto Rican hills at Arecibo had been filled with a giant concave radar dish wider than three football fields strung end to end. With its immensely powerful transmitter, the dish focused a pure-note radio beam toward Mercury when the planet was overhead. Like a ripple in a pond, the pulse spread out away from Earth, but traveling at the speed of light. A few minutes later, when the expanding shell-shaped beam swept past Mercury, the tiny portion of it that was intercepted by the planet was reflected back, just as a post sticking from a pond generates its own ripple when another ripple passes by. As the original pulse expanded infinitely toward the stars, Mercury's own weak echo sped backward through interplanetary space, again at the speed of light. Minutes later it flashed noiselessly past the Earth. But one radio receiver was pointed and tuned to receive the one-millionth of a trillionth part of Mercury's expanding echo-bubble that it intercepted—the same monstrous Arecibo dish that had emitted the original pulse a quarter-hour earlier. The power of the Arecibo transmitter is exceeded only by the sensitivity of its receiver. So exact was the reception of the infinitesimally weak echo that the radar astronomers working in Puerto Rico didn't merely detect it but could analyze bits of the echo to determine Mercury's rotation rate.

How does one measure spin rate from an echo? Imagine you are facing a cliff a mile away and you clap your hands. About 10 seconds later, if you listen carefully, you will hear the echo. Clap several times again, once a second. The 10-second delayed echoes faithfully return, once a second. But imagine that, rather improbably, the cliff is rushing toward you as you clap once a second. The first echo is delayed nearly the full 10 seconds, but the cliff is much closer by the time your next clap reaches it, so the echo returns on the heels of the first one. And the third echo returns even more quickly. So the echo frequency is much greater than the one-a-second clap frequency. Without even seeing the onrushing cliff, a person might apprehend the imminent catastrophe threatened by the peculiarly speeded-up echoes. Just as the frequency of echoes increases as the mountain advances, so the frequency, or pitch, of the echo of a pure note reflected from the advancing cliff would be raised.

Now imagine that, instead of a cliff, you stand before a fixed, but rapidly rotating, globe. If the globe spins clockwise, then its right-hand portions are rushing toward you, and its left-hand side is receding. The parts rushing toward or away from you, as the globe spins, are somewhat farther away than the front side of the globe that is moving right to left. Suppose you could isolate just the later echoes from the most distant parts of this globe. Those coming from the right-hand side would be raised in frequency, or pitch, just as from the onrushing cliff or like the whistle of an approaching locomotive. But the simultaneous echoes from the left-hand (receding) side would be lowered in pitch. The faster the globe spins, the greater would be the difference in pitch of the echoes from the approaching and receding parts.

Similarly, the pure-note pulse from the Arecibo radar was slightly modified in pitch by Mercury's spin. The spread in pitch found in Mercury's echo was greater than expected from a planet spinning once every 88 days. Probably the spin was somewhere between 54 and 64 days. More precise radar measurements have since pinpointed the value at between 58.4 and 58.9 days. The announcement of the radar results caused consternation among optical astronomers, who only then reanalyzed the older drawings and found the occasional discrepancies with the 88-day period. Measurements of the drawings and some improved telescopic photographs refined the period to 58.65 days, almost exactly two-thirds of the 88-day orbital period. Finally in March 1974 the Mariner 10 spacecraft hurtled past Mercury in an orbit that brought it back again six months later. Measurements of close-up photographs confirmed that, within a hundredth of a day, the rotation period is exactly two-thirds the orbital period.

Not until the two-thirds period was found did astrophysicists reexamine George Darwin's original theory to learn why Mercury fails to keep one face sunward. The answer lies in the same un-equal orbital velocities by which Schiaparelli tried to explain away apparent oscillations of Mercury's spots. When Mercury is closest to the sun in its elongated orbit, it zips around so fast that even the 59-day spin lags behind. So when Mercury is being tugged most strongly by the sun, it actually does keep one side more or less facing sunward. Those periodic pulls on Mercury's slightly bulging figure overcome the weaker attractions when Mercury is farther from the sun and hold it locked into a precise two-thirds spin. How often it is that theory lags behind empirical data!

Before the first spacecraft exploration, radar astronomers had learned still more about Mercury. By timing the echo delays, they measured Mercury's diameter to nearly one part in a thousand. Radar also helped to improve our estimate of Mercury's mass by measuring Mercury's influence on the positions of other planets. Dividing mass by volume, the radar astronomers calculated that Mercury is made of material denser than rocks from which our Earth is made. Presumably Mercury contains much iron, the only cosmically abundant material of high atomic weight. Mercury's high density had been suspected for decades from crude earlier data, but radar established it beyond doubt. Radar also provided an estimate of the roughness of Mercury's soil (rougher than Venus's) and demonstrated the lack of any Mercurian continental-scale highlands and lowlands, such as exist on Earth and Mars.

It is not often appreciated that ingenious improvements to ground-based optical-, radar-, and radiotelescopes have more than once enabled astronomers to do from the ground what experts had thought, a few years earlier, could be done only by the vastly more expensive examination of a planet by spacecraft. There has even been a tendency for the popular press, and uninformed spacecraft experimenters themselves, to credit the space program with discoveries made months or years earlier by ground-based astronomers. For instance, it is widely written that Mariner 4 first showed that Audouin Dollfus's estimate of the atmospheric pressure on Mars was too high by a factor of 10. Yet the year before Mariner 4 reached Mars, University of Arizona astronomers Gerard Kuiper and Toby Owen had lowered Dollfus's estimate by a factor of 5 and were close to deriving the currently accepted value. However, the dramatic feats of radar astronomy in the mid-1960s could not be overlooked. The unanticipated coup of reorienting our conception of Mercury gave the technique a tremendous boost in prestige. The attendant funding has enabled the Arecibo radar to be improved to the degree that it is now beginning to map the surface of Venus beneath that planet's thick clouds in nearly as fine detail as Mariner 10's beautiful photographs of Mercurian landscapes.

On February 3, 1970, scientists gathered at the California Institute of Technology to review what was then known about Mercury and to plan the exploration strategy for the Mariner spacecraft that was to be launched in November 1973. The National Aeronautics and Space Administration had already selected teams of scientists to build the various instruments that would study Mercury's magnetic field, its charged-particle environment, and its infrared and ultraviolet radiation, as well as the cameras that would transmit back the close-up pictures of Mercury's surface geology.

Imaging-camera experiments always seem to be first in NASA's scheme of spacecraft priorities. Some people feel we learn more about a planet from pictures than by studying it in unfamiliar wavelengths or measuring the planet's interaction with its inter-planetary space environment. But that is prejudice arising from our human biology—the superiority of our vision to our other senses. Instrument-design experts have overcome this bias, so in response to more worldly pressures they often vie with each other for specific spacecraft and mission designs that will benefit their own experiment. In the end the imaging experiment usually wins, however, for taxpayers like pictures and little understand or appre-ciate infrared radiation, electron spectra, magnetic field intensi-ties, or other technical measurements.

Yet on February 3, 1970, the imaging experiment was in trou-ble. Most scientists agreed that useful measurements could be obtained from virtually all other instruments only if Mariner flew past the dark side of Mercury. For instance, space physicists could hope to understand the interaction between Mercury's magnetic field and the wind of protons and electrons streaming away from the sun only by passing through Mercury's wake. Pictures could not be taken of the dark side, of course, and sunlit parts of the planet would be visible only from great distances long before or after encounter. From so far away the available television camera, identical to that flown to Mars on Mariner 9, would yield much fuzzier Mercury pictures than Earth-based telescopic views of the moon. Imaging team members used the Caltech gathering to rally support for a film (rather than television) camera system that would yield sharp pictures even from far away. (Some participants, noticing some film company representatives in attendance, thought the lobbying effort was actually the underlying purpose for the whole conference.)

Predictably, the choice of a camera system for the Mariner 10 mission had little to do with science and much to do with sociol-ogy. Geologists hoped to get general coverage of the sunlit half of Mercury, plus highly magnified views of at least the more interesting localities. But there were many other constraints. Only a finite number of pictures can be transmitted accurately back to Earth; some transmission capacity had to be reserved for other instruments on board. The film camera system advocated by the Imaging Science Team would require costs not anticipated in the mission budget, and it seemed risky to go with a system never tried before. Moreover, it was doubted that a film system would be as "photometrically accurate" as a television system. Scientists would often like to know not just that this place is darker than that one, but indeed that it is, say, just 17 percent darker; from such data they can measure how much the ground slopes or how big are the particles that compose the soils. Accurate calibration of photographs is difficult and television systems are ostensibly more accurate.

That was a sore point, however, for all previous television-equipped spacecraft had returned pretty pictures that nonetheless were photometrically worthless. The worst problem occurred in 1965, when Mariner 4 returned pictures of Mars that were horribly washed out and were made presentable only by extensive computer processing. Andrew Young, a brilliant, nonconformist expert on photometry, incurred the displeasure of some officials at the Jet Propulsion Laboratory by suggesting bluntly that Mariner 4's problem resulted from negligence. Stray light from Mars flooded the vidicon through a hole in the camera; the hole was a defect in the design, but nobody had bothered to check for such off-axis light leaks before installing the instrument in the spacecraft.

The arguments for the film system by Bruce Murray and the Imaging Team were impressive, yet there were counterarguments for frugality and conservatism. An eventual compromise enabled Mariner 10 to fly to Mercury equipped with the tested and true Mariner 9 television, but also with a larger telephoto lens that permitted sharp pictures to be taken from the great distances required by the dark-side trajectory. Additional technological advances further improved picture quality to nearly the predictions for the film system. Given that our previous best views of Mercury showed nothing more than Schiaparelli's vague, darkish spots, it was inevitable that any close-up imagery would have thrilled the photogeologists. So as encounter date approached, the imaging-system debate was forgotten. Even the 1970 Mercury conference itself faded from memory, and a subsequent conference held on the same Caltech campus in June 1975 was dubbed the First International Colloquium on Mercury.

By early afternoon on Friday, March 29, 1974, the level of excitement had reached a feverish pitch in the catacombs of the buildings on the Jet Propulsion Laboratory campus, nestled against the San Gabriel Mountains north of Los Angeles. Over 148 million kilometers away the Mariner 10 spacecraft was winging past Mercury at more than 11 kilometers per second, having been propelled there from its fling past Venus only 7 weeks earlier. Its signals were being received by a specially upgraded tracking antenna in the Mojave Desert and relayed in turn to the Mission Control Room at JPL. Television monitors scattered around the JPL campus flashed pictures of a landscape never before seen. Later, and into the night, further data streamed back to Earth from a spacecraft that was already leaving its target far behind. Measurements made while Mariner was behind Mercury, and thus invisible from Earth, had been recorded on tape and were played back shortly before sunrise on Saturday morning.

Overwhelmed by the flood of data and bleary-eyed from lack of sleep, scientists who had been working for years toward the success of these few short hours were straggling about, trying to comprehend what it all meant. Reporters waited on Sunday morning for a news conference to begin so they could report around the world on what had been learned about the innermost planet. Although a proper analysis of the data had not even begun, the press was impatient, so the scientists gathered on the stage, still doing last-minute calculations. They grinned at the success of their mission, but their faces also sagged with fatigue. One particularly sleepy experimenter, who had examined data from an instrument sensitive to ultraviolet rays, blurted out his discovery of a moon circling Mercury. He proceeded to name it after his family pet, a fact that was duly reported by the wire services. After he had slept a few nights, his mind cleared and he realized that the Mercurian moonlet was a mere background star. But by then the reporters had gone.

Finally, outside the public spotlight, the hard but rewarding work of serious analysis was to begin, analysis that would take months, perhaps years. After hiding in the sun's glare for an eternity, Mercury had at last been exposed to scrutiny.

6

The Inside View

As HUMAN BEINGS living on the surface of a planet, we are naturally most concerned about that surface: the land, the waters that flow on the land, and the air that blows across it. The outer reaches of our atmosphere produce the shimmering light shows we call the aurorae, but otherwise seem of little relevance. Although we are amused by Jules Verne's fantastic voyage to the center of the Earth, we derive economic benefit from only a very narrow zone of our planet, near its surface.

As we comprehend the complexity of modern civilization, we become more aware of the interrelation of ourselves and our greater environment. Our ecological concerns are no passing fad, but a critical stage in our uncertain progress toward understanding our environment so that we may control our ever-widening effects on it and ensure our survival. We are learning that the ozone layer in the stratosphere, far above the altitudes at which most aircraft fly, is essential to our survival, and yet it is endangered by our use of aerosol sprays. The geological readjustments deep in the Earth threaten the survival of our largest cities. Ice ages may result from a crucial interplay between processes far removed from the Earth's surface; mighty volcanic eruptions, which originate deep in our planet as drifting crustal plates plow beneath each other, throw volcanic dust high in the stratosphere where it is trapped for years, affecting the balance of solar energy.

In this chapter I discuss one remote part of a planet—the deep interior. But, paradoxically, measurements in that other remote zone, far above the surface and atmosphere, tell us most about the interiors of some planets. The interior of our own planet not only constitutes the vast bulk of its mass, but its influence on us is surprisingly great, so we must understand it. And comparisons of Earth's interior with interiors of other planets help test our ideas about our own planet.

Planetary interiors may seem so inaccessible as to be quite undecipherable. But that is not the case. Were Earth a totally static body with its interior never manifesting itself on the outside, we would know nothing but its volume calculated from the measured exterior dimensions. But the mass of the Earth's interior pulls on us from a distance—a fact as fundamental to our existence as the solar energy that sustains us. The life-supporting waters and vapors are thereby held close to the surface on which we too were permanently confined until the Space Age dawned. From this same essential gravitational pull we can measure the mass of the Earth. Since we know its volume, we can calculate the Earth's density, which is a fundamental constraint on the chemical composition of the interior.

We learn about some of the minerals that compose the Earth's interior from direct examination. I have on my desk a heavy, granular, greenish rock made of olivine crystals, a material believed to be a prime constituent of the Earth's mantle. It is no coincidence that I found this rock while walking on the slopes of Mount Hualalai on the island of Hawaii, one of the most active areas of volcanism. The veins of olivine-rich rocks in which diamonds are mined were formed deep in the Earth and were brought to the surface in several places by our planet's active geological forces.

The earthquakes that threaten us provide another way of exploring the Earth's interior. While the mighty shocks cause damage only in limited areas, seismographs sense weakened shocks from all over the world. From the delays between quakes and their detection around the globe, seismologists calculate propagation velocities that constrain the densities of the layers of the Earth's interior. The forms of the seismograms further reveal whether the material traversed by the wave is solid or liquid. It is an unambiguous result of seismology that the outer layer of the Earth's core is liquid; indeed it must be molten nickel-iron with some admixture of silicon or sulfur.

Yet another manifestation of the Earth's interior is our magnetic field. It is produced by the convective boiling of the Earth's liquid metallic core, although the actual process is not yet well understood. For centuries mankind's exploration of our planet's surface was guided by compasses that are directed by these subterranean forces. The field is so strong that some rocks have retained magnetization from previous epochs. It is from studies of such rocks that the paths of continents and sea floors have been traced around the globe. The recognition that the continents move leads us in turn to search for origins of their motions back in the bowels of our planet.

Deciphering the internal properties of the Earth from these and other surface manifestations is a tricky business. What progress has been made owes much to laboratory measurements of the properties of rocky materials at the high pressures and temperatures that characterize the Earth's interior. Laboratory techniques fail to reach the central pressures of nearly 4 million atmospheres and temperatures of over 4,000 degrees K (7,000 degrees F), so theoretical physics and cosmochemistry must provide still further insights into the total picture.

Planets are like living creatures. They are born, full of life and activity. They mature, consume energy, and settle into established ways. Finally they run down, become dormant, and die. On a human time scale planetary lives are virtually eternal. We see only a snapshot of each planet and can only surmise its evolution. True, we can measure (barely) the several-centimeter-per-year drift of the continents. But most of the Earth's dynamic activity that we can actually measure, such as the rate at which the land is washed into the sea (estimated from the sediment load of rivers), is superficial and transient. Such surface processes would soon stop were it not for the much longer-lived internal activity of our planet and of the sun. Yet the snapshot of our planet is filled with evolutionary clues, ranging from fossils to chemical isotope ratios. Moreover, well-understood laws of physics and chemistry enable us to predict generally the future evolution of the Earth.

What is fundamental to a planet's life history, as for the universe in general, is matter (of which it is composed) and energy (which keeps it moving, or "living"). At birth a planet has a certain amount of matter; maybe a lot, like the Earth, or less, like the moon. Planets also began with different temperatures; some originated near the newly formed sun and were hot, while others farther away were colder. (There were other contributions to the initial temperatures of the planets, which I have discussed before.)

Then the planets began to cool, to lose their heat. Heat can be moved in three ways. First, it can be directly radiated as light or infrared radiation. That is how one gets hot sitting before a fire, or sunburned by a star 93 million miles away. But just a sheet of cardboard shuts out light, so it is easy to believe that the Earth's interior hardly relies predominantly on radiation to move its heat out to the surface through thousands of kilometers of cold, opaque, solid rock. Radiation is of course an important way for surface heat to be transported up through the atmosphere and into space.

In opaque, solid materials heat flows by conduction between touching objects. This is how one is warmed by touching a warmer object. Heat slowly flowing from a warmer place to a colder place is a major way that planets cool off.

If temperature differences are great and the material is fluid, convection occurs. This is the actual transport of a warm hunk of material to a colder region. When the sun warms air near the ground on a hot summer day but the air above is very cold, the warm air is buoyed up into a billowing thundercloud and replaced by cold downdrafts at the edges of the storm. Over eons of time even the solid parts of the Earth's interior are plastically deformable and are slowly stirred by convection. Warmer parcels of the interior rise toward the surface and transport heat faster than by conduction alone. Some geophysicists believe that continents are conveyed on top of such slowly churning convective cells within the mantle. Hot lavas pouring out onto the surface from deep inside the Earth are an extreme example of convection and graphic proof that the Earth is still "alive."

A planet's lifetime depends on its size. Just as a glacier lasts longer in the sun than does an ice cube, so a larger planet takes longer to cool than a smaller one. Hence the Earth should be alive long after the smaller asteroids and moon have cooled off, everything else being equal. But planetary evolution is not quite that simple because the initial heat is not the only source of energy. For one thing, the sun shines and continues to heat planetary surfaces. Even more important is the heat released by the decay of radioactive isotopes over their half-lives of several billion years. These isotopes were not equally distributed among the planets because of the variable composition of matter condensed from the cooling solar nebula. Moreover, they are not uniformly distributed within planets; because of their chemical affinities, they tend to float toward the surface of a planet if it warms to the point at which iron begins to sink to the center.

Thus the fate of a planet depends not only on its size and original temperature, but also on the quantity and internal distribution of radioactives and on their half-lives. By using the best data on the thermal properties of rocks, geophysicists such as Sean Solomon of MIT have been able to calculate how different parts of a planet become warmer or cooler. When it becomes warm enough for rocks to melt and iron to sink, Solomon moves all heat sources in his computer model into the upper regions of the planet, to simulate the migration of radioactives into the crust. He then uses thermal parameters for metallic iron in his mathematical representation of the core and those for rocks elsewhere. Gradually, as the computer churns away, the heat is calculated to migrate to the surface of the planet, where Solomon's computer code lets it radiate away. Finally, programmed by their half-lives, the radioactive heat sources waste away. The computer then predicts the inevitable fate of each planet's heat engine. Unless Solomon has overlooked something terribly important, the Earth is predestined to cool and die several billion years from now.

The internal death of the Earth need not doom life on the Earth's surface, for our dominant energy is from the sun, which will still be shining, perhaps more brightly than ever. But the ecological effects of the Earth's internal death will be profound and are not easily predicted. Volcanism will stop and the continents will cease to drift. Mountains will no longer form and if our atmosphere is unchanged—a dubious proposition—the rains will wash the land into the sea. The eventual fate of Earth depends on the finite supply of fuel for the thermonuclear reactions within the sun. Ultimately the sun itself will die and the Earth will be a cold cinder drifting in the blackness of space.

For our knowledge of the Earth's interior we rely on indirect evidence, much of which has been accessible because of the extreme geological activity of our planet. The churning motions constantly signal our seismic stations, toss interior rocks onto the surface, and generate a strong magnetic field. How much more difficult it must seem to learn about the interiors of less active planets, or of the moon, whose ancient cratered surface testifies to the relative tranquillity of its interior. It is as though an amateur spy in a cheap motel were trying to learn about the occupants of neighboring rooms. From one room come shrieks of laughter, loud music, and the sounds of smashing glass. Perhaps a fist crashes through the thin dividing walls. Our supersleuth would have no difficulty guessing that a drunken party was in progress and, by listening to the raucous noise, might even learn the names and personalities of those attending. But from the other room there is no noise at all. Is the room occupied by a sleeping man? Or a woman? Or a dead body? Or is it a broom closet and not a bedroom at all?

The problem of learning about other planetary interiors is exacerbated by the brief and quite superficial nature of our exploratory ventures into space. The seismic stations on the moon, with which we listen to the occasional, weak moonquakes, are expensive and pitifully few. Not even a mechanical lander has been on Mercury. The two Soviet Venus explorers that took pictures from remote locations on that planet's forbidding surface in autumn 1975 survived the heat for only an hour. Thus we must study the relatively passive interiors of many terrestrial planets from afar, without benefit of touch.

Still, the advances in comprehension of the Earth, primarily in the two decades since the International Geophysical "Year" (July 1957–December 1958), have given geophysicists sufficient background to infer the nature of other planetary interiors given only data sensed remotely and returned from spacecraft flying past.

Consider again Mariner 10's reconnaissance of the planet Mercury in 1974. Newspapers printed photos of Mercury that reminded everyone, astrogeologists included, of the pockmarked surface of the moon. I have discussed in earlier chapters how these monotonous crater fields hold important implications for the rest of the solar system. Yet the most significant results from Mariner 10 are not primarily from the surface pictures but concern the interior of Mercury as it has been pieced together from a synthesis of several Mariner 10 experiments, theoretical calculations, and the few facts that were known before.

Even the most passive planetary interior reveals itself by its gravitational effect on the motions of neighboring planets and asteroids. So it was long suspected, and more recently known for sure, that Mercury is quite dense. Almost certainly it is composed of a large amount of iron, nearly two-thirds by mass. This contrasts with about 35 percent iron for the Earth. The ordinary chondrite types of stony meteorites, which are thought to be unmodified condensates from the primordial solar nebula, have only 20 to 25 percent iron.

A fundamental question is whether the iron is uniformly distributed throughout Mercury as it is in chondritic meteorites or is concentrated in a central metallic core as in the Earth. If the iron were uniformly distributed, then Mercury could be an unmodified, ancient aggregate of condensates that were simply more iron-rich than in the portion of the solar system where the chondrites formed. Even with an iron core, Mercury could be unmodified if it accreted so fast that it grew layer upon layer as successive minerals condensed from the more slowly cooling solar nebula, beginning with iron. According to the prevailing models of planetary formation, however, Mercury should have formed from homogeneous material, similar to (though more iron-rich than) the ordinary chondrite meteorites. If Mercury has a core, then the planet must have heated and partially melted, permitting the heavy iron to sink to the center of the planet. Such core formation, believed to have occurred on Earth early in its life, would virtually have turned the planet inside out and had a catastrophic effect on its surface.

Mariner 10 revealed evidence for the probable existence of a core in Mercury, but it also raised further questions. The most convincing evidence came, ironically, from instruments that measured the most superficial aspects of Mercury: the nature of interplanetary space near the planet. The three teams of scientists running the plasma science experiment, the magnetometer, and the charged-particle telescope were all surprised at the implications of their data: Mercury has a magnetic field! While it has less than 1 percent the strength of the Earth's magnetic field, it is still substantial, contrasting markedly with the virtual absence of fields around Mars and Venus.

The discovery was unexpected, for while a molten core is

thought to be a prerequisite for a planetary magnetic field, rapid rotation was thought to be required as well. Mercury's languishing 59-day spin hardly seemed to qualify, and Mariner researchers had expected to study Mercury's passive interaction with the interplanetary fields and associated charged solar particles streaming past the planet. Such passive interactions, carefully analyzed for the moon by a series of lunar satellites, are sometimes complex and superficially seem to mimic an active, intrinsic magnetic field. Thus, the experimenters were at first uncertain of their discovery. After all, there is an awfully large volume of space around Mercury and on March 29, 1974, Mariner 10 made only a single traverse of it.

But by a fortunate circumstance it was possible for Mariner 10 to pass through Mercury's field again a year later. Although not recognized when the Mariner Venus-Mercury mission was first planned, the spacecraft's path was a peculiar one that brought it back near Mercury every 6 months. During the second encounter slight course adjustments were made so that Mariner would pass high over sunlit parts of the planet, outside the magnetosphere, to get wide photographic coverage of the sunlit south polar regions that were poorly observed on the first pass. But there was a third encounter, in March 1975, which was once again aimed close to the night side of the planet, though closer to the pole than the first pass. Spacecraft instruments don't function forever; the third Mercury encounter proved the last for Mariner 10. Yet the third encounter provided a second cut through Mercury's magnetosphere that was sufficient to confirm the expectations from the first pass that Mercury indeed has an intrinsic magnetic field.

An intrinsic magnetic field isn't necessarily generated by a currently molten core, however. After all, a bar magnet doesn't have a molten center, but nobody has deduced how Mercury's rocks might have been sufficiently magnetized to produce the observed field. One current hypothesis is that, despite Mercury's slow rotation, its field is being generated in the same rather mysterious way that the Earth's field is. That implies not only that Mercury has an electrically conducting iron core, but also that the core is at least partially molten.

We could also infer the presence of a core from examining Mercury's surface composition. This might seem paradoxical at first, but if the iron required by Mercury's density were absent from the surface, then it must have sunk at least partway toward the core. In Chapter 4 I described how Earth-based astronomers measure the composition of surface rocks on asteroids and other distant bodies. For Mercury, however, the techniques fail to provide a clear clue to the percentage of iron present.

An indirect clue about surface composition comes from interpretation of the television pictures of Mercury's broad, more lightly cratered plains. In most respects they resemble the moon's lightly cratered volcanic maria, so Mariner 10 geologists think they too were once composed of molten rock. And iron will surely sink beneath the areas where molten rock originates. Expanses of volcanic plains certainly would suggest that Mercury segregated into a metallic core and rocky mantle. But this nifty argument for core formation is less convincing than it seems. It bears an uncanny resemblance to arguments proposed several years ago that some bright highland plains on the moon were volcanic. When Apollo 16 astronauts brought back rocks from such a plain, they turned out to be impact-smashed crustal rocks and were not volcanic at all, to the embarrassment of United States Geological Survey geologists who had confidently mapped the bright volcanic plains. Perhaps the Mercurian plains aren't volcanic either and say nothing about a core within that planet.

Another trait of Mercury's topography provides more secure evidence for a molten stage and core formation on the planet. One feature on Mercury is virtually absent from the moon: the so-called lobate scarps, or cliffs, which are scattered all over Mercury and range from tens to hundreds of kilometers in length. Several are shown in the picture on page 45. They have the special shapes and forms of particular scarps on the Earth that form when immense compressional forces thrust one block of land on top of another.

In fact, Mercury's crust resembles rather closely the wrinkled skin of a piece of fruit that has dried out and shrunk inside.

What could have caused Mercury's interior to shrink, leaving a crumpled surface? The only plausible answer is that the interior was once warmer and contracted upon cooling. Indeed, Sean Solomon has calculated that the cooling of normal rocky materials in a Mercurian mantle alone would account for the shrinking indicated by the size and number of lobate scarps on Mercury. He has further shown that if Mercury's core had also cooled and solidified, much greater contraction could have resulted, crumpling Mercury's surface much more than is observed. He concludes that Mercury's core must still be molten, as is suggested also, of course, by the presence of the magnetic field. Since some large craters have been wrinkled and yet other craters formed on top of some wrinkles, the contraction of the mantle must have coincided roughly with the period of heavy cratering, presumably a long time ago. The core has yet to contract and it is still molten.

The logic seems obvious. Yet if it is correct, there must be something dreadfully wrong with our understanding of the chemical composition of planets, or of the thermal evolution of planets, or with the cratering chronology for Mercury. Solomon has used his computer programs to trace the evolution of Mercury's temperature, using current hypotheses for the chemical composition of Mercury and sources for the heat that has melted it. The surprising answer is that, in the absence of some additional source of heat early in Mercury's history, Mercury's expected complement of radioactive uranium, thorium, and potassium should not have heated the planet sufficiently for the iron to sink inward for $1\frac{1}{2}$ billion years. That is long past the popularly hypothesized cratering cataclysm. Yet once the core formed, it must have cooled rapidly, according to Solomon's computer runs, and should have solidified long before now. So the straightforward model yields a core that forms too late to explain the wrinkled craters, yet solidifies too early to explain the magnetic field and lack of still more wrinkles.

I have mentioned in other chapters the independent evidence afforded by meteorites and the asteroids for an early source of heat for those bodies besides long-lived radioactive decay. Such early heating of Mercury would permit its core to have been formed earlier. Thus the association of wrinkles and craters may provide a further clue that early heating was a widespread occurrence in the solar system and not a peculiarity of the evolution of just one, or a few, planets.

More difficult to explain is the failure of Mercury's core to solidify. There must be a large continuing heat source in Mercury's interior. Possibly the repetitive stretching and compression of the planet by the ever-changing solar tides raised in it as it swings close to the sun and away again contribute some heat. Or perhaps Mercury has more radioactive isotopes decaying within it than prevailing cosmochemical models predict. Maybe we even misunderstand the overall time scales for Mercury's development as inferred from craters and their relationships to the great wrinkled cliffs.

The study of Mercury's interior exemplifies the impetus given to planetary science by new observations. Theoreticians who worry about how magnetic fields are produced in planets must rethink their models that depended upon a rapid planetary spin. They must ask afresh why slowly spinning bodies like Venus and the moon lack such fields. Chemists are spurred on by the challenging possibility that their models for Mercury's composition may be more imperfect than they realized. And so on. Scientists who had never thought about Mercury at all may now find their esoteric laboratory or theoretical work to be the key to rendering compatible the seemingly conflicting constraints.

Even our perception of the Earth's interior and its magnetic field may require revision. We cannot foresee how important, or unimportant, these changes may be for our day-to-day lives. One example illustrates the potential importance for our past. Rocks that record the strength and direction of the Earth's magnetic field show that it changes directions every quarter of a million years or so. The north magnetic pole is in the Antarctic, and then it flips back. It is not known how long it takes the field to flip, only that it is much shorter than the duration between flips. These traumatic events in our planet's magnetic life could be very long compared with a human lifetime, and should the magnetic field somehow turn off during flips, there could be dire consequences. After all, our magnetosphere controls the influx of charged particles from interplanetary and intergalactic space. There could be modifications of our lower atmosphere and dangers for living things if this protective regime were disrupted. Indeed, there are hints in the fossil record that mass extinctions of certain species may have coincided with magnetic field reversals. On the other hand, a thorough understanding of these processes, achieved through comparisons with magnetic fields of other planets, may demonstrate that the flips are quite harmless.

7

The Vapors of Venus
and Other Gassy Envelopes

BREATHE DEEPLY. Think of the fluid envelopes that surround the Earth—the air and the seas. Their miraculous properties shape the variegated environments that have provided the ingredients for the origin and sustenance of life. Without the atmosphere and the hydrosphere the Earth's surface would be very simple, despite all the subterranean churnings. Cratering impacts and occasional quakes would shake the ground, as on the moon, and only gravity would pull molten or fragmented rock downhill. The minerals would be few and life processes impossible.

But with gaseous and liquid coatings the Earth's surface is transformed into a veritable chemistry lab of cycling reactants and end products. The fluidity permits rapid migrations; sediments and raindrops sink, bubbles and light gases rise, animals move about. Only partially transparent, the air and water absorb some of the sun's energy, transporting it far and wide in currents and storms. Powerful sparks flash through the sky, redressing charge imbalances. Rains flush the ground, exposing fresh rock to chemical attack. Massive deposits of sediments accumulate on the sea bottoms, eventually to be cycled through the Earth's high-pressure oven below and spewed forth again from volcanoes.

This rich, wet, and gaseous chemical feast made the origin of life feasible, indeed probable, on our planet. If the processes that created our envelopes were unique to Earth, other rocky worlds would be monotonously moonlike. The possibilities of life on them would be remote. Yet visual astronomers of the nineteenth century never doubted the Earth-like traits of our neighbors in space. They had good evidence for vapors on the other worlds. They couldn't appreciate, from watching dust clouds drift across Mars's surface or scrutinizing the cloudy veil of Venus, how terribly un-Earth-like those extraterrestrial atmospheres were.

Astronomers have long known that Venus has an atmosphere. That was not an inference just from the brilliant and nearly featureless expanse of white that always seems to shroud the planet. When Venus passes nearly between Earth and the sun—closer to us than any other planet comes—its thin crescent extends into a ring of light, entirely surrounding the planet. There can be no doubt: we are witnessing both dawn and twilight on our sister planet as sunlight is refracted through a substantial·atmosphere. Beneath the shroud Venus has hidden her secrets from us and remained more mysterious than planets ten times more remote. Through brute force, however, modern technology has finally enabled us to penetrate the veil.

To understand Venus's atmosphere is to understand Venus and much about Earth as well. Venus is nearly the same size and density as Earth and is not that much closer to the sun. We say the barometric pressure is high at 30.5 and the temperature hot at 98 degrees F (37 degrees C or 310 degrees K), while on our twin planet normal conditions are a pressure of 3,000 and a temperature of 890 degrees F (750 degrees K). Venus's clouds of sulfuric acid blanket a choking ocean of carbon dioxide, a hundred times as dense as our own atmosphere. At night the ground glows dully from the extreme heat; the days are gloomy beneath perpetual clouds.

Imaginative fiction writers and scientists alike found Venus's hidden surface fertile ground for speculation. Astrophysicist Fred Hoyle imagined oceans of oil on Venus that would have satiated our energy needs. C. S. Lewis portrayed a watery world. The first hint about conditions on Venus came in the mid-1950s when radio astronomers detected strong emission from Venus. It took a while to prove the emission was due to a hot surface, rather than to something else. Subsequently, ground-based radar echoes from Venus showed that its day is very long, 243 days, and that it spins from east to west, rather than in the direction of the Earth and most other planets. Radar probing determined the diameter of the solid part of the planet and, in combination with other data, showed that the cloud tops are about 60 kilometers above the surface. As the power and sensitivity of radar observatories have improved, we have gotten increasingly detailed maps of canyons and other topography far beneath Venus's clouds.

Meanwhile, a barrage of spacecraft have reconnoitered Venus. While American Mariners have been content to study it from afar, the Soviet Union has been dropping instruments into its atmosphere. The early Soviet experiments frequently quit operating prematurely, providing dramatic evidence of the unexpectedly hostile environment near the surface. In October 1975 the Russians achieved a major first: they radioed back pictures from the surface of Venus, showing plains strewn with oddly shaped rocks. Although they succumbed to the heat after an hour, Veneras 9 and 10 first clocked winds of 2 to 8 miles per hour in the soupy air. Above the cloud deck the American Mariners have found that Venus lacks a magnetic field. More recently, excellent pictures of Venus's cloud patterns were taken by Mariner 10 on its way to Mercury.

Even the Earth-based astronomers have lately contributed to our understanding of the structure and composition of Venus's uppermost layers. At last we have enough data in hand so that chemists, physicists, and mathematicians can employ their brain power (computerized and natural) to understand how and why the Venusian atmosphere behaves as it does. There are even emerging some intelligent speculations about why the volatile envelopes of the twin planets, Venus and Earth, have evolved so differently.

Why is Venus so hot? It is not because it is near the sun. Mercury is closer yet cooler. Actually Venus should be comfortably cool. Even a small telescope reveals the brilliance of Venus's clouds; so much sunlight is reflected that little must be absorbed. Just as a house with a white roof and walls is much cooler on a sunny day than a black-shingled house, so Venus might be expected to be cool beneath its clouds.

The origin of the planet's heat has been controversial for a decade and a half, the clouds themselves figuring prominently in the debate. Unless Venus were nearly molten due to heat generated within the planet—and that seems most unlikely— the clouds must behave differently from white paint. According to Carl Sagan, they must behave like the walls of a greenhouse. To visualize the greenhouse effect, imagine a greenhouse made of glass. Glass is transparent to light, so the sun shines through and heats up the ground and plants inside. They in turn radiate the heat back. But whereas the sun's 6,000-degree-K surface radiates primarily visible light (it is no accident that our eyes are tuned to the sun's wavelengths), the reradiation by the room-temperature plants occurs at much longer, infrared wavelengths. Glass is opaque to infrared radiation, absorbs it, and reradiates it back to the plants again. Since a little heat does leak out, because glass is not perfectly opaque to thermal radiation and also conducts heat, the temperature in the greenhouse doesn't rise continually higher.

It was recently pointed out to embarrassed meteorologists, who have debated the relevance to Venus of their greenhouse calculations, that this effect may not even be important for greenhouses, however. Outside ground warmed by the sun heats adjacent air, which then floats upward to where the barometric pressure is less. The air parcel expands, cools, and settles into equilibrium. Meanwhile, at the ground the warmed air is replaced by cooler parcels from above. This process of stirring or convection, mentioned earlier in the book, warms upper regions and keeps the air near the ground from getting too hot. Air on the Earth begins to convect whenever the temperature begins to drop with altitude more quickly than about $6\frac{1}{4}$ degrees C per kilometer. So except in an inversion, when the upper air is relatively warm, convection maintains the $6\frac{1}{4}$-degree-C-per-kilometer temperature profile, which is why mountaintops are cool. The reason it is warmer inside than outside a greenhouse is mainly that the roof keeps the warmed-up inside air from floating away by convection. The greenhouse effect, although a perfectly valid physical principle, contributes only some of the additional heat inside a greenhouse.

There is no lid on Venus and the dense carbon dioxide is free to convect. The temperature profile of Venus's atmosphere, inferred from Mariner fly-bys and measured directly by Venera landers, ranges from 750 degrees K at the ground to 250 degrees at the cloud tops. That is just the 8 degrees C per kilometer expected for a convecting atmosphere of nearly pure CO_2. Can the combination of 100 Earth atmospheres of CO_2 plus the clouds plus whatever else is there serve as the hypothetical greenhouse glass? Is the Venusian atmosphere sufficiently transparent in the visible spectrum and opaque in the infrared to account for the furnacelike heat?

The small amount of carbon dioxide in our own air absorbs some radiation, making it difficult for astronomers to measure stars at some infrared wavelengths. The vast amount of CO_2 on Venus would absorb still more, but there remain holes in CO_2's infrared spectrum allowing some radiation to leak through, keeping the temperatures from rising. Another absorber is necessary to maintain the heat. In the 1960s Carl Sagan advocated water in the form of vapor and ice clouds as the missing absorber. Although Earth-based spectra revealed little measurable water above Venus's clouds, Sagan calculated that a layer of water-ice clouds, with a humid atmosphere beneath, yielded sufficient infrared absorption to make the Venus greenhouse work.

Beyond being the preeminent popularizer of astronomy on late-night television shows, Carl Sagan has been the most dynamic and influential planetary scientist since the early 1960s. A man of vivid imagination, he keeps alive a wide variety of conceptions of planetary environments. By suggesting often outlandish alternatives and challenging traditionalists to disprove them, he has inspired doubts about many accepted theories. Sagan's role is essential for a healthy science because a bandwagon effect frequently leads to premature consensus among scientists before equally plausible alternatives have even been thought of, let alone rationally rejected. Carl Sagan regards his advocacy of unusual hypotheses as an intellectual sport, designed to keep his colleagues on their toes, and he claims to be dispassionate about the actual truth of his models.

But Sagan has not always taken a detached view of the Venusian clouds, which have been a major part of his serious research since his days as a graduate student. Long after most of his colleagues agreed that his once-accepted water-ice model for the clouds of Venus was incompatible with radio and polarimetric data, the loquacious Sagan continued to press his case. Sagan watchers were forced to conclude that he actually believed in water-ice clouds (as most scientists believe in their own theories). Some even began to wonder if he believed also in his whimsical gas-bag creatures on Jupiter or polar bears on Mars!

Then, in March 1973, at a meeting of the Division for Planetary Sciences of the American Astronomical Society, all the confusing data about Venus's cloudy veil suddenly fell into place. In a remarkable set of talks in a crowded auditorium in

Tucson's Executive Inn, three scientists reported independent research that suggested the clouds were made of droplets of sulfuric acid. The first to think of sulfuric acid seems to have been Godfrey Sill, who works at the Lunar and Planetary Laboratory in Tucson. His arguments for sulfuric acid were bolstered by those of Jim Pollack, of NASA's Ames Research Center near Palo Alto, and of Andy Young. Several years later the arguments still seem persuasive.

Yet one cannot forget Carl Sagan's water entirely. The clouds, although highly concentrated (75 to 85 percent acid), are after all a water solution of sulfuric acid. The vapor pressure of water above such an acid cloud would be very low, which accounts for the small amounts measured spectroscopically from Earth. Below the cloud bottom, wherever that may be, there could be much more water, although not too much; otherwise it would have been detected by radio telescopes.

A boost to Sagan's case came when Russian space scientists reported that the chemistry experiments on some Venera landers measured quite a lot of water. Few planetary astronomers have seen the inside of a chemistry lab since they were college freshmen, so it may have been hard for them to evaluate the Russian result. Michael Belton, an astronomer at the Kitt Peak National Observatory, decided to reeducate his colleagues during an October 1974 Venus conference at the august Goddard Institute for Space Studies in New York City. Belton explained how one Russian gas analyzer was designed to measure CO_2. Venusian air was let into a double chamber, separated by a pressure-sensitive membrane. One chamber contained a pellet of potassium hydroxide, which absorbs carbon dioxide. After the absorbing chemical had done its work, the pressure between the two chambers indicated how much of the Venus air is CO_2. Belton, apologizing for his failures as a high school chemistry student, produced a beaker of sulfuric acid. He reminded his colleagues that Venera sampled air from within the sulfuric acid cloud, which of course was unknown when the Russians designed their spacecraft. With a mischievous grin he remarked, "I don't think this has ever been done in the annals of planetary astronomy. Let's see what happens when we put potassium hydroxide in the sulfuric acid."

The published conference proceedings record the result as "Fizzle. Fume. Fizzle." The Goddard audience was more impressed. They left the meeting, no doubt, with little faith in the Russian analyses of the composition of the Venusian air. But Belton's dramatic point may have been a bit unfair. The Venusian clouds, after all, are not like a beaker of sulfuric acid; rather they form a tenuous, smoggy haze. In fact, during one conference session Belton's associate Don Hunten tucked in his chin and proclaimed that "this stuff [Venusian cloud] is actually more transparent than the air outside that window." Indeed sulfuric acid contributes to the eye-stinging quality of smog in cities where high-sulfur fuels are burned, so the analogy between the Venus clouds and smog is not an entirely casual one.

The atmosphere of Venus seems remote from our daily concerns. Occasionally planetary studies have direct relevance for us, however. Alternatives to Sagan's discredited water clouds, before sulfur, included certain chlorinated chemicals. Several planetary physicists made theoretical studies of the behavior of chlorine compounds in the upper atmosphere of Venus. Their research now has little relevance for Venus, but it hit home when it was realized that chlorinated aerosol-spray propellants rise into our upper atmosphere and consume the protective ozone layer.

The best guess now is that the combination of carbon dioxide and a sulfuric acid haze on Venus is sufficiently opaque in the infrared to block in the small amount of sunlight that filters down to the planet's surface. Perhaps water vapor in the lower Venusian atmosphere helps as well. Light-measuring devices on the Soviet landers, especially the cameras on Veneras 9 and 10, prove that sunlight really does reach the surface. The Russians had been sufficiently dubious on this point that they equipped the latest Veneras with floodlights to illuminate the scene for the cameras.

They proved unnecessary. So the greenhouse effect seems to do a lot for Venus, although it would be unfair to say that the debates are entirely settled. If one can't even be sure the greenhouse effect operates in greenhouses, one shouldn't be too dogmatic about Venus.

That is what we think the Venusian atmosphere is made of and why it is hot. But what is the weather like on Venus? Are there winds and storms that make Venus's atmosphere a participant in the activity and evolution of the planet? On Earth we depend on static properties of our air—the oxygen and the temperate climate. But equally essential for us are rains, alternating with sunny days, and the winds that blow, scattering seeds and dispersing smog. We must also understand such damaging weather phenomena as hurricanes, hailstorms, and droughts.

It may seem presumptuous to analyze the weather on Venus when our own is so poorly understood. Yet it is for very practical reasons that we must study Venus. The failures of the weather bureau to warn us of last week's blizzard or of the thunderstorm that washed out the company picnic result from the lack of theoretical understanding of instabilities in fluid flows. These problems have been attacked by ingenious mathematicians faced with extremely complex equations and too few data from the only laboratory at hand, the Earth's atmosphere. Fast and powerful computers can precisely model the evolution of storms, but these numerical storms develop more slowly than the real storms outside the window. Another terrible complication in weather prediction is that our atmosphere is so unstable that the tiniest swirl of leaves can grow into a mighty storm in just a couple of weeks. It would be impossible to have a network of weather stations that could keep track of every such whirlwind.

In order to simplify the equations and make shortcuts for the computers, it is necessary for meteorologists to learn better how and why the atmosphere churns about as it does. They know that density is important, as are the vertical temperature profile and gravity. The heating is fundamental since it drives the atmospheric "engine." Also crucial is the rate at which a planet spins. Fluid dynamicists think they understand how some of these factors are related to the large-scale circulation. They have tried to test their ideas by dropping dyes into rotating dishpans of water and tracing the motions of the dyes. But the Earth is vastly larger, and there are many characteristics of its oceans and atmospheres which can hardly be modeled in a laboratory. Just imagine how you would make and hold together a free spherical shell of fluid in a lab!

So Venus, Mars, Jupiter, and other enveloped planets are natural laboratories for testing theories of the general circulation of our own atmosphere. The atmosphere of Venus is much thicker than ours, that of Mars much thinner. The air on Mars is warmed directly by the sunlit ground (except during planetwide dust storms), while Jupiter's clouds are warmed both from the outside by the sun and from the inside by the giant planet's own internal heat. Venus spins much more slowly than the Earth, Jupiter much faster. Mars has less gravity than the Earth, Venus the same, and Jupiter much more. A whole generation of theoretical meteorologists are testing their models for terrestrial meteorology by applying them to the vast quantity of spacecraft data from these other worlds.

Nearly two and a half centuries ago English meteorologist George Hadley presented a paper before the Royal Society of London in which he proposed an explanation for the trade winds. The sun heats the tropics much more than the poles, he noted, yet the poles are not all that cold. Hadley suggested that tropical air rises and flows toward the poles, forming a circulating cell with an undertow of cool polar air returning to the tropics at low levels. This "Hadley cell" formed a major element in models of the Earth's circulation long after it was realized that twisting of flows by the fast rotation of the Earth disrupted single-celled transport of heat between the tropics and the poles.

In the 1970s it is becoming clear that Hadley's picture of

the Earth's circulation may be just right for slowly rotating Venus! It was first proposed in the mid-1960s that solar heat absorbed in the clouds might be transported in a single cell away and down from the portion of Venus directly under the sun. It was hoped that such motions might produce the sizzling ground temperatures without having to rely upon the then-controversial greenhouse effect. Downward heat convection is no longer required to explain the surface heating, which is fortunate since it is hard to imagine warm air sinking. Furthermore, Venus's atmosphere is so thick that it cannot respond to the daily radiation of the sun, even though "daily" for Venus means several Earth months. Rather, Venus's dense atmosphere senses the sun's warmth as a cover surrounding the planet, more intense at the equator than at the poles. Theorists now expect that there are traditional Hadley cell winds, with greenhouse-warmed air rising in the Venusian tropics. The Hadley model is consistent with what few data we have about winds and temperatures underneath Venus's clouds, but it remains a hypothesis in search of confirmation.

The visible clouds constitute only the tiniest, most tenuous part of Venus's upper atmosphere. Their motions may be influenced by the underlying winds, but they do not exhibit the motions expected for the top of the Hadley cell. We should not be surprised, however, since the winds in the Earth's stratosphere and above differ greatly from those we measure at the ground.

The first indication of cloud motions on Venus came 50 years ago from photographs taken through filters that admit only ultraviolet light. Unlike normal photographs of the blank cloud deck, ultraviolet pictures show changing patterns of dark patches on Venus. Not until the 1960s was it finally recognized that the patches move about Venus once each 4 Earth days, 60 times faster than the planet itself turns, or at a speed of 110 meters per second (250 miles per hour).

Are there really such strong winds in the stratosphere of Venus?

Common sense may suggest that the rushing cloud patterns reveal how fast the winds must be blowing. But think a bit about familiar motions, real and apparent. Tap a taut rope and a wave rushes along the rope; yet the rope doesn't move horizontally. As you bob up and down in ocean waves, look at the water: while the waves sweep relentlessly to shore, the water stays pretty much where it is. Perhaps you have watched clouds pass overhead and seen a clear patch approaching; but the clear patch never makes it as clouds continuously form and thicken at its leading edge. Or consider a cloud suspended over a mountaintop while a stiff breeze blows right through it; evidently the cloud is not a wind marker like a balloon, but marks the position of a so-called standing wave as the air flows over the topographic obstacle.

Mike Belton has been studying Venusian cloud motions from the beautiful ultraviolet close-up photographs taken by Mariner 10. The largest global features are horizontal, Y-shaped dark patches very reminiscent of the ultraviolet features photographed from Earth. He has concluded that the moving patches represent wave propagation, not the actual motion of Venus's air. Yet the waves apparently move with respect to the air at only a few tens of meters per second, much more slowly than the air itself moves around Venus. Just as the motion of a lantern on a flatcar is due more to the speeding train than to the walking man carrying the lamp, so most of the 4-day apparent circulation of Venus is real wind motion after all, and not the illusory motions of Belton's waves.

Mariner 10 pictures show long, narrow belts near the equator of Venus that gradually shift southward. Where the sun is roughly overhead, there are cellular cloud structures, some dark, others bright, inside. Near this subsolar region there often are long, tilted streaks that resemble bow waves around a ship, or shock waves generated by an obstacle in supersonic flow. Belton has found terrestrial analogs for all these features. Some were recognized on Earth only since satellites began photographing our own planet regularly. The cells, he believes, are caused by convection under conditions analogous to shallow maritime inversion layers in the Atlantic and Pacific, which generate similar cloud patterns. The convection disrupts the rapidly flowing circulation of air around Venus, generating the bow waves. Mike Belton and his colleagues have measured the dimensions, motions, and lifetimes of all these wave phenomena. They can then infer the temperatures and pressures in the upper atmosphere of Venus and learn still more about that planet's weather.

There remains one major puzzle, however. What are the dark ultraviolet spots? By analogy with weather satellite pictures of Earth they might seem to be holes in the clouds, but spectroscopic studies show that they are at the same altitude as the bright clouds. Furthermore, the sulfuric acid clouds have properties more like a smoggy haze layer than discrete clouds. Rather small temperature variations cause terrestrial water vapor to condense or volatize readily when the humidity is high. But sulfuric acid is very nonvolatile, so what manifests the propagating waves on Venus? There would seem to be something in the clouds the thermodynamic behavior of which enables minor changes in temperature to change it from a clear gas to an opaque cloud. Godfrey Sill suggests that bromine dissolved in hydrobromic acid might have such properties. Others have suggested that pure sulfur, which is dark in the ultraviolet, might be responsible. Nobody finds any proposal especially convincing, and it almost seems simpler to imagine that some cosmic artist is dabbing ultraviolet paint on some of the clouds of Venus!

Venus, the Earth, and Mars are all similar planets, although Mars is somewhat smaller. Theory suggests all were made of roughly the same chemical mix of materials. All have probably formed an iron core. And all three planets absorb roughly the same amount of solar energy; Venus, though closer to the sun, reflects away a greater percentage of sunlight while Mars, which is farther, absorbs more. We are challenged, then, to understand why the atmosphere of Venus is 100 times as dense as our own and why that of Mars is more than 100 times less dense than ours.

Atmospheres are not remnants of primeval blankets that surrounded protoplanets. Rather, the Earth's atmosphere, and probably others, has leaked out from the depths over the eons. Why are gases exuded from these chemically similar planets now so different? The answer lies in chemical interactions, which, in combination with physical, geological, and biological processes, have greatly exaggerated the minor differences among the three planets.

It seems that Venus was just enough warmer than Earth to cause an irreversible sequence of events that boiled away its oceans and led to the present inferno. On Earth, water remained stable and provided a rich environment for the origin of life; life, once established, completely changed the chemistry of the air and water to the conditions familiar today. Mars is a bit colder and smaller than Earth, so some of its volatiles have frozen into underground and polar ice deposits while others have leaked off the top of the atmosphere into space because of the weak Martian gravity.

Atmospheric chemicals are prone to react with each other and with the ground until everything is in chemical balance and no further reactions occur. But the Earth's atmosphere is in a state of constant chemical activity, never reaching equilibrium, as new rocks are rapidly produced by volcanic and mountain-building processes; the winds and rains wash away the chemically broken-down or weathered rock, always exposing fresh rocks to the air. Radar images of mountain ridges and canyons suggest there has been great geological activity on Venus, just as is certainly true of some Martian provinces. But there are no oceans on Mars or Venus in which to deposit sediments. There are not even rains or streams to wash away soils, though there were on Mars in times past. Sulfuric acid rains fall through the corrosive Venusian clouds, but they never reach the ground. Perhaps winds on Mars and Venus are sufficient to scour away soils continually so that their atmospheres eventually may approach chemical equilibrium with rocks throughout the crustal layers of the planets. If not, reactions involving just the surface layer of a planet may have had only a superficial effect on the atmospheric chemistry. Even the great heat on Venus, which is conducive to rapid chemical reactions, could not overcome an absence of erosion processes, and the atmosphere and planet would be chemically isolated.

Among the processes that cause planetary atmospheres to depart from chemical balance are things that occur at their tops. If the temperatures are high enough, or reactions provide sufficient "kicks," atoms may be accelerated beyond escape velocity. Over the eons great quantities of the lighter gases may evaporate away into space as they are replaced by upward diffusion from below. A particularly potent influence on upper atmospheric gases is ultraviolet solar radiation, which is sufficiently energetic to tear a water molecule into its constituent parts, hydrogen and oxygen. Hydrogen, the lightest gas, evaporates rapidly. Sunlight can also photodissociate other gases. For instance, carbon dioxide is converted to carbon monoxide plus oxygen, which are then available for further reactions. The steady-state composition of an atmosphere depends on the rates of disequilibrating processes relative to reaction rates and relative to rates at which winds mix the atmospheric gases.

A major disequilibrating process is life itself, which has had a profound effect on our own air. Originally the Earth's atmosphere was much richer in carbon dioxide. It lacked oxygen, including the rich form we call ozone. Unshielded solar radiation may have provided the energy necessary for synthesis of life in the early unoxidized oceans. Eventually oxygen began to accumulate in the Earth's atmosphere, first from the photodissociation of water, then from the photosynthesis of carbon dioxide by living plant cells.

Oxygen, carbon dioxide, and water, being in chemical disequilibrium with rocks, help to weather rocks into soils that are washed into the sea. Some of these in turn are used by ocean life to form shells which, along with chemically precipitated carbonates, accumulate into great deposits of limestone and other carbonate rocks on the ocean floors. At continental margins the sea-floor-spreading treadmill carries the rocks down into the Earth's hot mantle. From the descending plates, molten lavas rise again to the surface, disgorging great quantities of carbon dioxide and other gases which, having reached chemical equilibrium with the interior of the Earth, are out of equilibrium with the surface and once again begin to eat away at the rocks.

The quantity of carbon contained in carbonate rocks in the Earth's crust is roughly the same as that contained in the carbon dioxide in Venus's massive atmosphere. That may mean that all the carbon dioxide that has leaked out of Venus's interior remains trapped in the atmosphere and has no way of participating in a geochemical cycle with the Venusian crust the way it does on Earth. Alternatively, at the broiling temperatures at the surface of Venus, 100 atmospheres of carbon dioxide may be the equilibrium value (we can't know for lack of good evidence on the minerals that compose Venus's rocks). But the high temperatures on Venus are partly due to the vast atmosphere of CO_2, so the question remains as to how it ever got that way to begin with.

The answer seems to lie in the evolution of water on Venus. Venus probably had its fair share of water initially and thus had a water-rich and carbon dioxide–rich atmosphere like the primitive Earth's. But evidently Venus was just near enough to the sun, hence sufficiently warmer, that the water created a runaway greenhouse. The infrared absorption of water vapor helped to warm the young Venus. This vaporized even more water, which trapped still further heat. Eventually the steamy atmosphere, containing an ocean's worth of vaporized water, approached today's temperatures. Ultraviolet sunlight then broke down the water vapor into hydrogen, which immediately escaped, and oxygen. The fate of the oxygen is uncertain; it might have escaped as well, or perhaps it oxidized crustal rocks. Anyway, once Venus got so hot there was no way for it to cool again.

The story for Mars is very different and somewhat uncertain. Both water and carbon dioxide, greatly exceeding the amounts now present in the tenuous Martian atmosphere, are stored in the polar ice caps. During summer much of the CO_2 reenters the atmosphere and is deposited at the other pole as dry ice. Much greater quantities of ice are probably trapped and frozen in the uppermost layers of the Martian ground. Volatiles exuded into the atmosphere by the volcanoes, or by occasional melting of some ice, were photodissociated by the sun. Certainly much water vapor has been destroyed this way, with hydrogen evaporating into space. As with Venus, the fate of the oxygen is uncertain; it may have been lost at the top of the atmosphere, too, or it may have helped oxidize the iron in Martian soils, producing the rusty color of the planet.

The widespread dry river channels on Mars testify to an earlier epoch when the atmosphere was different and waters flowed. If the crater erosion and channelization occurred early in the history of Mars, before a thick early atmosphere evaporated into space, we might account for the rivers as a once-only event in Martian history. Alternatively, and more interestingly, the wet period may have occurred rather recently, during a temporarily warmer climate; the ice thawed and generated an Earth-like atmosphere for a time, before cooling temperatures froze the ice again to await another future warming.

Although Venus seems to be pretty much trapped in its present hot, acidic state, the atmospheres of both Earth and Mars will continue to evolve. With the arrival of technological man on the scene, that evolution may be very rapid, indeed, for until man evolved, the biosphere and atmosphere of the Earth lived in a harmonious, symbiotic chemical relationship. Now man has introduced great disequilibrating changes that may, or may not, drastically alter the traits of our life-sustaining envelope. If some theorists are correct, the Martian atmosphere may be even more susceptible to human influence—for good or ill. Before anyone tries to start runaway atmospheric evolution on Mars, we should all consider the merits of designating Mars an international wilderness area to preserve for eternity the history imprinted in its rocks of a more Earth-like time when Martian life might have thrived.

ABOVE: Scientist-astronaut Harrison H. Schmitt examines a large split boulder lying near the Taurus-Littrow landing site of the Apollo 17 Lunar Module. Astronaut Eugene A. Cernan took the picture. *Courtesy NASA*

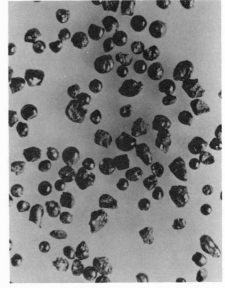

RIGHT: The rocks and soils brought back from the moon by the Apollo astronauts have been a gold mine of information for lunar scientists. This is a highly magnified view of some of the most enigmatic lunar material: the so-called orange soil discovered by astronaut Jack Schmitt near Shorty Crater, at the Apollo 17 landing site. The particles are actually the size of fine sand and may be of volcanic origin. *Courtesy NASA*

The cloud-bedecked planet Earth rises over the moon's horizon, as viewed from Apollo 17's lunar orbit, shortly before it returned to Earth in December 1972. No more lunar expeditions are planned for at least a decade. *Courtesy NASA*

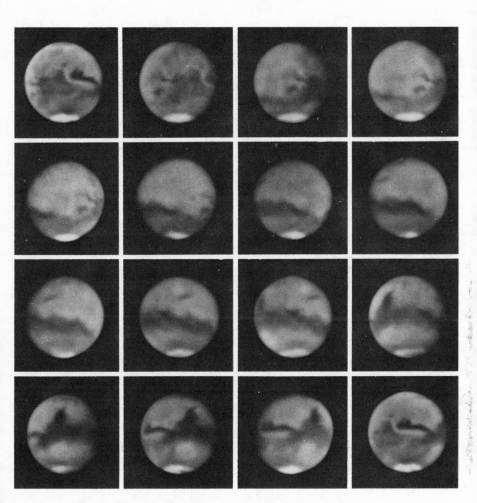

As Mariner 9 sped between Earth and Mars in 1971, this is how Mars looked from Earth. This sequence of photographs, taken by the International Planetary Patrol Program coordinated through Lowell Observatory, is arranged left to right and top to bottom. As Mars spins around once every 24 hours and 37 minutes, new features are continually brought into view. The prominent dark patch north of the equator in the lower left picture is Syrtis Major, which was first seen by the Dutch scientist Christian Huygens in 1659, not long after the invention of the telescope. *Courtesy W. Baum*

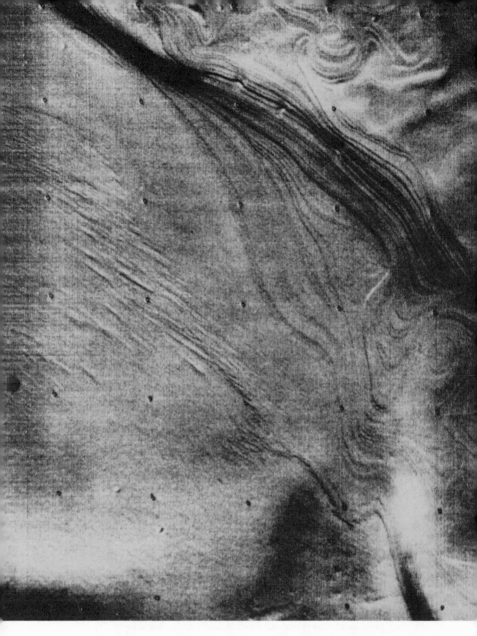

Streaks and swirls dominate this view of an 88-by-63-kilometer portion of the south polar region of Mars. Evidently there is a sequence of thin layers of deposits piled on top of each other. What kind of cyclical process might deposit or erode such layers is not yet understood. *Courtesy NASA*

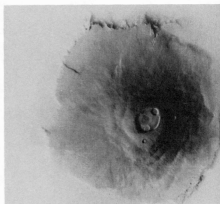

The Olympus Mons shield volcano on Mars is the largest volcano known on any planet. First named Nix Olympica ("Snows of Olympus") because of its frequent white appearance as viewed through telescopes from Earth, its true nature is hinted at in photographs taken as Mariner 6 approached Mars in 1969 (above left). Nix Olympica is the whitish ring with darker interior just above the middle of the planet. Perhaps a hint of the summit crater can been seen, but the hints went unrecognized until the 1971 dust storm dropped beneath the flanks of the huge volcano, revealing the impressive feature, 500 kilometers in diameter, to Mariner 9 (above right). In the high-resolution view of the texture of the slopes of Olympus Mons (left; downslope is to the upper left), the trained eye of a photogeologist quickly recognizes the ridges, cracks, and flowing tongues of material as characteristic of lava flows on the sides of Earth's volcanoes, such as Mauna Loa in Hawaii. *Courtesy NASA*

ABOVE: An eroded network of gullies has shaped the northwest flanks of the Alba Patera volcano. Whereas underground sources of water may explain large channels elsewhere on Mars, these seem to have required a general rainfall, which suggests that the climate of Mars was very different in the past. Most of the gullies visible here are about $\frac{1}{2}$ kilometer across. *Courtesy NASA*

BELOW: This close-up view of a Martian river channel system looks remarkably like an aerial view of a river valley on Earth—except there is no water now flowing on Mars. The picture was taken by the narrow-angle Mariner 9 camera from a distance of about 1,800 kilometers.

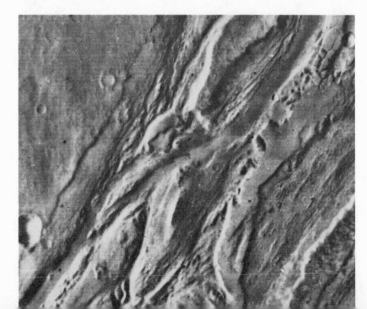

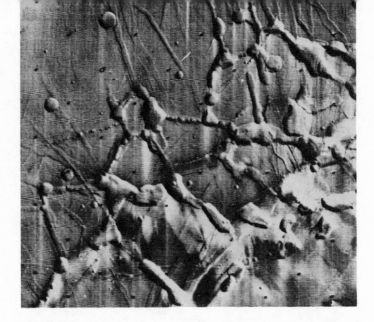

ABOVE: This complex region of grooves, crater chains, and canyons lies at the western end of the great Valles Marineris canyon complex, which stretches for thousands of kilometers to the east. It may mark the early stages of canyon formation as the giant rift progresses westward. This region of Mars is 400 kilometers across. *Courtesy NASA*

BELOW: This small section of the Valles Marineris is 440 kilometers long. Extensive erosion of the canyon walls is evident. Note the chain of craters dividing the two canyons. *Courtesy NASA*

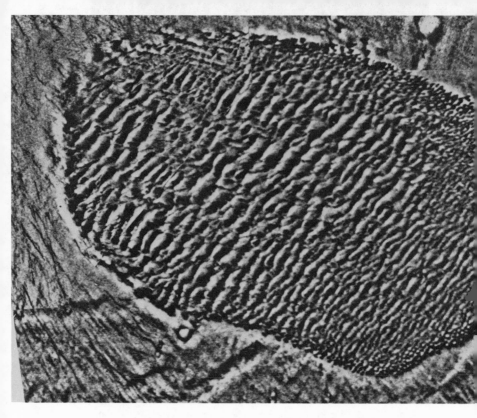

ABOVE: Many Martian craters have dark patches on their bottoms. This close-up view of one such patch reveals it to be a great field of black sand dunes measuring 65 by 130 kilometers. *Courtesy NASA*

BELOW: Wind abrasion was probably responsible for the erosion that left these pedestal craters perched above a denuded plain on Mars. Evidently the blanket of rocks ejected from the craters during their formation formed a hard layer of ground that proved more difficult to erode than the surrounding terrain, leaving the craters as mesalike remnants. To the right is a double crater, which was probably formed by the simultaneous impact of two great meteoroids. *Courtesy NASA*

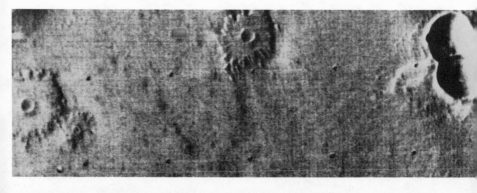

ABOVE: This broken, channelized terrain lies at the boundary between the two geologic hemispheres of Mars: the ancient cratered uplands to the south and the younger plains to the north. Erosion has progressed southward, leaving mesalike remnants of the uplands. This Mariner 9 view spans several hundred kilometers. *Courtesy NASA*

BELOW: Craters on the Martian surface disturb the atmospheric flow, causing wave patterns in the Martian clouds, highlighted by the late afternoon sun. The picture was taken on February 14, 1972, by Mariner 9. *Courtesy NASA*

BOTTOM: This panorama of sand dunes, rocks, and boulders on Mars was taken by the Viking 1 lander about two hours after Martian sunrise. The boom that supports Viking's small weather station cuts through the center of the picture. *Courtesy NASA*

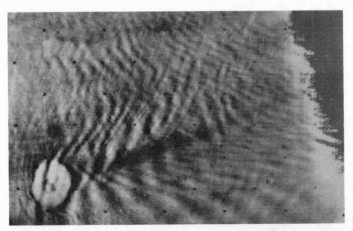

ABOVE: A panoramic view to the south-east taken from the surface of Mars by Viking 1. *Courtesy NASA*

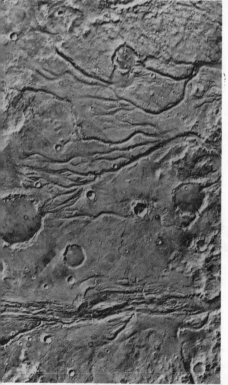

LEFT: Viking 1 Orbiter photographed this region on Mars, which apparently has been carved by large river channels. The ground slopes from west to east with a drop of about 3 kilometers across this picture. The Lunae Planum highlands are to the west and the Chryse plains, in which Viking 1 landed, are to the east well beyond the right-hand border. *Courtesy NASA*

RIGHT: This oblique view across Mars toward the southern horizon some 19,000 kilometers away was obtained on the night of July 11, 1976, by one of Viking 1 Orbiter's two TV cameras. The picture is dominated by the greatly eroded rim of the huge Argyre Basin, the smooth floor of which is seen to the left. Above the horizon are layers of haze, 25 to 40 kilometers high, probably composed of crystals of frozen carbon dioxide. *Courtesy NASA*

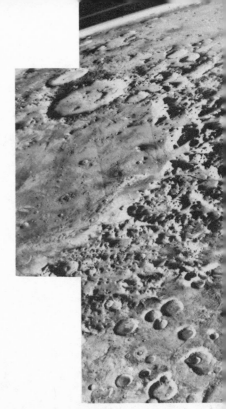

BELOW: The surface of Mars is covered with rocks in the vicinity of the second Viking lander. Many of the stones have small holes, or "vesicles," in them. The angular rock in the right foreground is about 25 centimeters (10 inches) across. A small depression or channel, possibly formed by flowing water, is visible near the top of the picture. *Courtesy NASA*

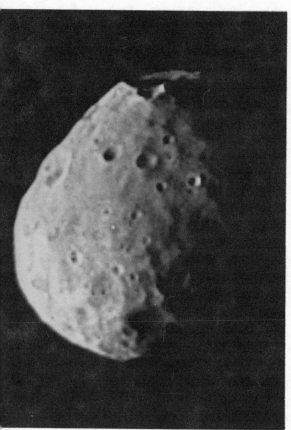

ABOVE LEFT: Viking 1 Orbiter snapped this view of Phobos, the inner and larger of two moons that orbit Mars. *Courtesy NASA*

ABOVE RIGHT: On September 18, 1976, Viking 2 Orbiter obtained this most detailed photograph ever of part of Phobos. Craters had been photographed on Phobos before by Mariner 9 as well as Viking 1 Orbiter, but scientists were flabbergasted by the numerous grooves—each roughly as wide as a football field is long—carved across the surface, almost as though a giant leaf-rake had been drawn across it. *Courtesy NASA*

8

The Moon: What Did We Learn from Apollo?

The termination of the Apollo flight program after Apollo 17 leaves the scientific tasks undertaken by Apollo substantially unfinished.

—REPORT OF THE LUNAR SCIENCE INSTITUTE, 1972

A MANSION STANDS on the brown, marshy coastal plains of eastern Texas, where the cultures of the American West and South mingle. A proud relic of bygone days, its balustraded patio overlooks Clear Lake, an estuary of Galveston Bay. Beyond formal gardens, now in disrepair, lies a tree-encircled private pond of several acres. A chandelier descends from the double-storied ceiling of the central hall and a broad, double staircase sweeps down toward the intricately carved fireplace. In these comfortable quarters, once the heart of the ranch of "Diamond Jim" West, lunar scientists from around the world gather to meet and talk and try to fathom the results of the greatest scientific enterprise in the history of mankind.

Emerging from an evening seminar in the Lunar Science Institute, few conferees notice the yellow orb rising between the branches, glowing dully through the sultry Houston smog. For them, the meaning of the Apollo Project and man's conquest of

Earth's satellite lies in the collection of exotic moon rocks, stored a mile away in one of the many buildings that compose the Johnson Space Center. As the years of lunar exploration recede into the past, the measurements and analyses continue, fostered and guided by the government-supported, university-controlled institute which occupies the restored West mansion.

Nobody's going to the moon anymore. For most Americans there are fading memories of Mission Control, a golf game amid the craters, and parachutes splashing in the Pacific. Maybe they recall questions of how old the moon was, whether it was hot or cold, and whether the meteor craters were actually volcanoes. And how did the moon form, and why? Perhaps people now take for granted that the answers were learned, although the news never made headlines. The media by then had turned to Cambodia, the fuel shortage, and Watergate, leaving Apollo scientists to study their moon rocks outside the limelight in which they had briefly been.

Taxpayers probably don't realize they have been spending upward of $15 million a year to learn about the nature and origin of the moon. But that abundance of wealth, envied by practitioners of less privileged sciences, is but a dribble compared to the flood of expenditures in the 1960s that brought the moon rocks back. As the dribble slows to a trickle, and one by one the lunar researchers return to studying meteorites and ocean basalts, the time has come to ask, "What have we learned?"

Did we find those pristine rocks from the earliest years of the solar system? No, we didn't. Is the moon hot or cold? It's warm. Were the craters formed by impacts or volcanism? Pre-Apollo opinions of the majority favoring impact haven't been changed, nor those of a few dissenters. How was the moon formed? Well, we still don't know. Then was the Apollo program uproarious, spectacular, and expensive, but yielding few results, and of no ultimate benefit to the taxpayers? I think such a conclusion very wrong. For the scientists who congregate at the Lunar Science Institute, the exploration of the moon has been a terrific success.

The benefits of Apollo are spreading throughout the physical and geological sciences. That we didn't learn immediately why the Earth has a moon is but a trifling detail to satisfied researchers. As for taxpayers, we must all stand back and examine the Apollo project and its relationship to science in fair perspective.

Why did we go to the moon? It is a simple question, but the answers are complex, involving politics, psychology, economics, and even foreign policy. It is a cliché, but well worth repeating, that the reason we went to the moon was certainly *not* to do scientific studies of the origin of the moon. Science, which never has been more than a tiny facet of human culture, would never by itself have rated the expenditures of tens of billions of dollars. At best, science occasionally rides piggyback on larger endeavors; such was true of Apollo.

What motivated Congress to vote the funds to go to the moon? That is the crucial question, because a necessary and probably sufficient requirement for going to the moon was the allocation of funds. Here is why Congressmen said they voted the funds: "The United States has not embarked upon its formidable program of space exploration in order to make or perpetuate a gigantic astronautic boondoggle. There are good reasons for this program. But, in essence, they all boil down to the fact that the program is expected to produce a number of highly valuable payoffs. It not only is expected to do so, it is doing so right now. . . . Those already showing up . . . include the most urgent and precious of all commodities—national security."*

Whether national security was prime mover or not, it is most unlikely that science was. Once the money was allocated, however, one might have expected a scientific orientation from the National Aeronautics and Space Administration, which administered the program. But even within the one small branch of NASA that was

*From *The Practical Values of Space Exploration,* Report of the Committee on Science and Astronautics, U.S. House of Representatives, Rept. No. 2091, 86th Congress, 2d Session, 1960.

responsible for science, other motives predominated in the early years of Apollo. In 1963 the then-director of the NASA Office of Space Science wrote: "While science plays an important role in lunar exploration, it was never intended to be the primary objective of that project. The impetus of the lunar program is derived from its place in the long-range U.S. program for exploration of the solar system. The heart of that program is man in space, the extension of man's control over his physical environment. The science and technology of space flight are ancillary developments which support the main thrust of manned exploration, while at the same time they bring valuable returns to our economy and our culture. . . . Thus the pace of the program must be set not by the measured patterns of scientific research, but by the urgencies of the response to national challenge."

Even today, with NASA's budget about half of what it once was, "man in space" remains the overall purpose of the space program. There is a powerful emotional, even spiritual, impetus behind man in space. It is captured in the novels of Arthur C. Clarke and it springs from the same motivations, endemic to Western culture, that have always impelled us to conquer the wilderness and physically transform it. Or one could be more cynical and suggest that NASA is like any other federal agency. Its power depends on its cash flow, and expensive hardware needed to shoot men into space is most important, not inexpensive science.

NASA administrators, and some scientists too, offer many rationalizations as to why the best way to study the solar system is to send human beings out there. But the arguments ring hollow, except to the true believer. Nowadays, space scientists watch the giant Space Shuttle project gobble up the bulk of NASA's allocations from Congress while once-in-a-century opportunities to study Uranus and Neptune are bypassed. These scientists lobby, ineffectually, for a reorientation of NASA's priorities that almost certainly will never come.

Other scientists, perhaps more practical and certainly more

attuned to political realities than their more idealistic colleagues, support the shuttle and the man-in-space activities. They believe that the little money left over for science from such enormous technological projects, and the occasional piggyback opportunities, will provide an adequate base for scientific research. If man-in-space projects were threatened, NASA might collapse and space scientists would be left holding an empty bag.

Our headlong rush to the moon was not a scientific research program and the astronauts (except for one) were not scientists. Once Neil Armstrong had taken his "one small step for a man" and the Americans had clearly beaten the Russians, public interest in moon landings nosedived and the series of subsequent Apollo missions was cut short. Now NASA can't even get the money to support one unmanned, polar-orbiting lunar satellite, designed to follow up on Apollo discoveries.

But if science wasn't important for Apollo, Apollo nonetheless revolutionized solar system science. The exhilaration of lunar scientists during the heyday of Apollo can never be recaptured. Theories debated for centuries were being resolved every week and hundreds of new mysteries were being revealed. Seismographs on the lunar surface were telemetering back totally unexpected reverberations deep below ground level. Lunar soils seemed unaccountably to have ages hundreds of millions of years older than the rocks from which they were derived by meteoritic erosion. The amount of data returned from the moon during the $3\frac{1}{2}$ short years from Apollo 11 to Apollo 17, both carried back in sample bags and radioed directly back, was truly staggering for scientists accustomed to waiting months for one small meteorite to fall from the skies.

Lunar science has now matured. The initial production-line measurements have all been done. The thoughtful, measured pace of scientific research has returned. There is now time for contemplation and for ingenuity to invent new techniques to unravel the history locked up in the lunar sample collection. The bright, imaginative physicists and chemists lured into the Apollo project now have their sophisticated instruments built and working, and reliable measurements are being reported and confirmed by other laboratories. Although the excitement of sample acquisition is gone, the serious process of experiment and synthesis continues, and new insights to the origin of the Earth and the moon are being reported at every meeting convened at the Lunar Science Institute.

The pre–Space Age picture of the moon was very incomplete. Its interior was a complete mystery. We could only observe radiation reflected by, or emitted from, the front side of the moon. The few clues we had about the moon's composition were indirect and inconclusive. We had pictures chiefly of its craters and plains. As augmented by intensified telescopic observation in the 1960s and the exploratory Ranger, Orbiter, and Surveyor spacecraft, our hazy pre-Apollo hypotheses for lunar evolution mainly concerned its large-scale surface geology.

The details in the moon's story have always been uncertain. Half a billion years after the moon's formation 4.6 billion years ago there was a torrential bombardment of the moon. Catastrophic impacts excavated great basins, throwing debris far across the moon and scouring preexisting craters. Subsequently the basins were flooded by lava flows, which cooled into overlapping layers of dark basaltic rocks. In the mid-1960s William Hartmann, then at the University of Arizona, estimated the period of volcanism occurred about 3.6 billion years ago, or 1 billion years after the moon had formed; we now know it started at least 3.9 billion years ago and ended about 3.2 billion years ago. There was occasional faulting of the lunar crust, but there were never even the beginnings of Earth-like crustal plate motions. During the last 3 billion years the moon has been struck and cratered by countless small bodies, and a few rather large ones, but otherwise its surface has been geologically dead.

These pre-Apollo ideas have generally survived the more extensive photography by the orbiting Command Module pilots and the close-up examination of several locales by astronauts on the surface of the moon. Of course, intelligent and imaginative scientists are adept at incorporating new data into their preconceived hypotheses; it is harder to be sure of the truth. Also, with so many data to assimilate in the short time elapsed since Apollo, scientists often accept uncritically the results of others outside their own narrow experimental or theoretical specialties. As the inconsistencies are gradually ironed out, some accepted interpretations of lunar geological history are sure to change.

While our views of lunar geology have only been sharpened by Apollo, entirely new and fundamental sides of the moon's personality have been revealed. These have come mainly from studies of the chemicals (and their isotopes) and the minerals of which moon rocks are made. We now know what the rocks are made of, and we think we know their origins. Since all moon rocks seem once to have been molten, we can deduce something of the thermal history of the moon. Some rocks must have been derived from others at great depth in the moon; just as the shape of a jigsaw puzzle piece reveals the shapes of its neighbors, the compositions of surface rocks provide clues about the lunar interior. The nature of the interior is further revealed by the way it affects the electrical and magnetic fields of the solar wind and in the way it propagates heat and seismic waves. The ages of moon rocks have been measured. Each rock records several ages, some reflecting the origin of the moon, others the gross separation of the moon into layers, and still others the subsequent volcanic and impact episodes that affected the different lunar provinces.

All rocks on Earth are composed entirely of minerals, each of which is a substance having a specific crystalline structure and chemical composition. We now know that moon rocks are made from some of the same minerals familiar in terrestrial and meteoritic rocks. That was no surprise to geochemists, who had never expected to find green cheese or any other exotic substance, but had no previous idea about which of those minerals to expect and in what proportions. The way in which chemical elements form mineral compounds reveals chemical abundances of the precursor material and the subsequent temperatures and pressures to which it has been subjected.

A common pre-Apollo idea was that the moon was made of chondritic meteorites, which contain the same proportions of most chemicals as the sun itself. As I have described earlier, the chondritic meteorites are the unaltered and unmelted materials that condensed from the original nebula in at least one part of the young solar system. Had the moon formed cold from chondrites and not been melted since, moon rocks would be chondrites. They aren't. All the stable chemical elements occur in lunar rocks, if only in trace amounts of parts per billion. Compared with chondrites, lunar rocks grossly lack some elements by factors of more than 1,000, yet are enriched in some others by factors of 10 to 30. Either the moon was not made out of chondrites or the chemicals have gotten rearranged on the moon so that the surface rocks are not representative of the whole moon. Or both.

It might seem presumptuous to pick up rocks from the moon's surface and claim to know what the bulk of the moon is made of. But geochemists have long studied the Earth's rocks and have developed pretty good ideas about what kinds of processes enrich certain chemicals and deplete others. For instance, as hot gas of solar composition cools, the first minerals to condense when the gas "cools" to 1,500 degrees K are calcium-rich compounds, followed by metallic iron, then the more common silicates, with water appearing relatively late (at the boiling point of water). So if the moon contained just those minerals that condense above a certain temperature, but none of the lower-temperature compounds, we could conclude that it condensed from a hot part of the solar nebula and has not been modified since. No other known chemical process would select chemicals in just the same way. The moon actually is enriched in many of the high-temperature compounds and depleted in some of the more volatile ones, but that is by no means the whole story.

Gold and nickel are absent on the moon, despite these elements' moderately high condensation temperatures. Since gold and nickel have a strong affinity for becoming alloyed with iron, it was immediately suggested that surface rocks are depleted of these minerals because they sank into a metallic core when the moon was molten. Yet we know from the lunar gravity field that a massive metallic core simply does not exist in the moon. It seems that the separation of iron-related elements from the moon occurred *before* the moon was made. I will return to just how the moon was made later, but for now we can accept that the moon was never chondritic but is enriched in high-temperature elements and depleted of volatile and iron-related elements.

There are still further ways in which the chemistry of moon rocks distinguishes them from meteorites, from terrestrial rocks, or even from each other. Most of these characteristics are very well known to geochemists familiar with rocks that have once been molten. For example, there is a group of chemicals known as the rare-earth elements which, because of their atomic sizes and poor ability to unite chemically with other elements, are poorly incorporated in mineral crystals. Thus when rocks are heated only a certain amount, a portion of the material melts into a liquid that is rich in rare-earth elements. Being composed of many different chemicals and minerals, rocks do not have a single melting temperature; rather an increasingly large portion of the less compatible constituents in a rock enter the melt as its temperature is raised over a range of hundreds of degrees. The rocks from the lunar maria are rich in such incompatible elements as the rare earths, strongly indicating that they are resolidified lavas, originally derived by the partial melting of deep interior rocks. Indeed, just as one can tell what meat was cooked by tasting the gravy, geochemists can infer the composition of parent rocks by analyzing the melts.

One rare-earth element, europium, acts differently from the rest. It seems to have been strongly depleted from the basaltic lavas that flooded the lunar maria basins. The depletion is probably due to europium's strong affinity for feldspar, a mineral rich in aluminum and calcium. It happens that in contrast to the maria rocks the lunar highland rocks are rich in feldspar and have an excess of europium. When molten rocks are cooled, experiments show that feldspar crystals are the first to form and tend to float to the surface. Their predominance in highland rocks, even those excavated from great depths by cratering impacts, suggests that the europium-enriched highlands formed from the top of a cooling ocean of molten rock, or magma. Evidently the europium-depleted maria lavas came later from the partial remelting of rocks that had settled toward the bottom of the original lunar magma ocean. Thus the complementary character of europium and other trace elements in maria and highland rocks reflects a period ending about 200 million years after the formation of the moon, when at least the outer layers of the moon melted and a gross chemical separation occurred.

How deep was the magma ocean? And from what depths were the lavas subsequently derived 3.2 to 3.9 billion years ago? These questions are central to our understanding of the moon, and there are some tentative answers. Velocities of seismic waves, caused by both tiny natural moonquakes and an occasional rocket or meteoroid striking the moon, indicate that the feldspar-rich highland rocks underlie both the highlands and the maria to a depth of about 60 kilometers. A magma ocean with an upper crust 60 kilometers thick must have been very deep indeed! That fact is confirmed by the presence of great mass concentrations, or "mascons," that pull on spacecraft orbiting the moon. The mascons seem to be the thick, dense, solidified lava lakes that fill the circular mare basins. They are so heavy that they could not be supported by the deformable, partially molten rock from which the lavas were derived. Were the partly molten layers close to the surface, the mascons would have sunk into the moon and would not remain today. The mascons must have been supported by a cold, rigid shell or crust well over 100 kilometers thick, so the lavas came from at least 100 kilometers down. But they cannot have come from much deeper than 400 kilometers, where the pressure is about 20,000 Earth atmospheres. The only kind of rock at such enormous pressures that could produce melts of lunar basaltic composition would be rich in garnet gems and have a density far greater than is permitted by measurements of the lunar gravity field. So the lavas came from depths of 100 to 400 kilometers, and 400 kilometers evidently marks the bottom of the magma ocean.

The lunar lava rocks differ somewhat in composition. Those found at the Apollo 11 and 17 sites contain a lot of titanium, while those at the Apollo 12 and 15 sites are titanium-poor. Since the former are ½ billion years older than the latter, geochemists have looked for ways in which the source region might have changed with time. Perhaps the temperatures of the parent rocks changed, producing melts of different composition. Alternatively, different parent rocks—perhaps at different depths—might have melted at different times.

There is a plausible scenario for the evolution of the lunar lavas that is consistent with what we imagine to be the thermal evolution of the moon. Since the moon is so small, compared with the other rocky worlds, it cannot retain its interior heat for long. So it cooled, especially at the surface. From the presence of the mascons we know that the upper 100 kilometers of the magma ocean had cooled within the first 700 million years. Presumably the outer layers continued to cool, and what heat was left migrated inward, where it was insulated by the thick lunar crust. The heat pulse eventually reached the core of the moon, which is evidently still quite warm today. The evidence for that is that a particular type of seismic wave that cannot move through liquids does not pass through the core of the moon; so the core is at least partially molten and well above 1,300 degrees K.

The effect of the heat pulse on melting rocks is the purview of experimental petrologists. These are scientists who subject rocks to high pressures and temperatures and study resulting changes in the minerals of the rocks, including the compositions of melts. Based on their experiments, they have developed models for the cooling of the early magma ocean that would explain the sequence of lava compositions. They believe that the rocks that could sweat out the high titanium lavas would be sandwiched at intermediate depths in the moon—below the feldspar-rich crust but above the kind of rocks that would give rise to low-titanium basalts. So as the heat pulse slowly propagated downward in the early moon, there were first high-titanium flows, then low-titanium ones, consistent with the measured ages of the rocks. Finally, 3 billion years ago, no lava at all could penetrate the thickening lunar crust, all lunar volcanism ceased, and the lunar surface became inactive.

It is generally true of lunar science that the facts and possibilities are so numerous that no simple model is entirely adequate. So it is with this scenario for the generation of lavas. Basalt titanium content has been inferred from telescopic measurements for regions never visited by the astronauts. The technique, described in Chapter 3, has been tested for regions visited by the astronauts and seems to work. Yet a crater-counting age-dating technique suggests that some of the apparently most titanium-rich regions of the moon are not old but rather are among the youngest, younger than any rocks brought back by the astronauts. So either the telescopic measurements are wrong, the age-dating technique is wrong, or the models for generating lavas at great depth in the moon are wrong.

In order to help resolve these questions, the Lunar Science Institute convened a conference in November 1975 entitled "Origins of Mare Basalts and Their Implications for Lunar Evolution." The ingenuity of the participants gave birth to numerous alternative models to explain mare basalts. Some theorists proposed that the basalts were derived from unmodified, primitive lunar material at great depths in the moon, well below the early magma ocean. Others found ways to produce lavas from very shallow depths in the moon. Even more exotic mechanisms have been invoked to explain the local concentration of highly radioactive rocks called KREEP (rocks that are rich in potassium [K], Rare Earth Elements, and Phosphorus).

In October 1976 the Lunar Science Institute initiated a large project, involving nearly 100 scientists, to compare lunar volcanism with terrestrial volcanism and with what we know of volcanism on Mercury, Mars, and even some asteroids. Geologists and astronomers who have never been interested in the moon before were enlisted in the hope that comparative studies, spurred by the exquisite and exhaustive work that has already been done on lunar volcanism, would lead to new insights about volcanism on all the planets, including the Earth. This fundamental stage of planetary evolution occurred early in the lives of such small bodies as some asteroids and even the moon, but it is at its peak right now on our own planet. Because of Earth's melting and chemical separation processes, analogous to those that occurred on the moon, we have available the rich concentrations of minerals called ores that have made possible much of human technology. It is hard to say whether or not there will be immense practical applications from such scientific understanding that can help recoup the vast expenditures of Apollo. But the comparative planetological approach being fostered by the NASA Office of Space Science through the Lunar Science Institute is clearly a first step in reaping the practical long-term benefits from learning about the moon. It remains for our country's uncertain political processes to determine whether or not there will be continued funds to support this important research.

Long, long ago, it is told, two Eskimo children were playing in their igloo during the eternal darkness of winter. The sister once chased her brother out into the darkness and they ran across the ground carrying their torches above them. Suddenly they were lifted into the sky, where they chase each other to this day. The girl is the sun. Her brother's torch is dimmer; he is the moon.

Modern science has had its own mythologies about the origin of the moon, some as quaint as the Eskimo legend. Some have said that the moon was wandering through the heavens from distant places when it came near Earth. The Earth reached out and captured it, forcing it always to face its captor. Others say the young Earth spun faster and faster as it formed until a huge chunk was flung out from the Pacific Ocean basin. As Eve was created from Adam's rib, so the moon was born of the Earth. Still others have imagined a twin birth in the Earth's orbit that yielded a double planet from the very beginning.

The wealth of data from Apollo has enabled the natural philosophers to embroider their simple tales to the point that they are stunning creations of human imagination. But as they all suffer from incompatibility with the laws of physics, or with the trace element abundances, or with other immutable facts, we can be sure that the truth has yet to be perceived. Scientists are optimistic, however, that from all the new data, combined with further study of other celestial examples, such as moonless Venus and moon-laden Jupiter, we may soon learn for sure how it is our planet has a companion. And perhaps, even now, we can begin to see the glimmerings of the final answer.

I have already recounted how the moon rocks have been depleted in metals like nickel and gold, which have a strong affinity for metallic iron and which would accompany iron into the core of a molten planet. In that sense the gross mineralogy of the moon resembles the crust of the Earth. Perhaps the moon was yanked from the Earth, with its iron-related elements already sunk into the core of the Earth. However, physicists think there are insuperable difficulties in getting a moon to escape as a whole body from the Earth's crust and evolve to its present orbit, leaving the Earth spinning at its present 24-hour rate. In addition, there are great differences between the Earth's mantle and the moon in terms of the abundances of such volatile trace elements as bismuth, sodium, and gold. These differences would be difficult to explain by known chemical processes.

It seems that the moon must have been made out of materials somewhat different from those that composed the primordial Earth. Some theorists even proposed that the moon formed near the planet Mercury from minerals that condensed first from the solar nebula at very high temperatures. That would account for the enrichment of those minerals in the moon, but ad hoc arguments were required to explain why the moon wasn't also enriched in iron, which is also a very high-temperature condensate. Furthermore, the celestial billiard-ball wizardry necessary to get the moon accelerated away from Mercury and then decelerated into a nice circular orbit around the Earth was more than any celestial mechanician was willing to vouch for. Even capture of the moon from less exotic locations of the solar system is not easy to imagine. A body hurtling past the Earth would zoom away again, not go into orbit. The Earth's gravity would tug at the retreating moon but would slow it down by no more than the amount it had accelerated it on its approach. An object coasting down a hill gains enough speed to carry it up the next hill; if it gets trapped in the intermediate valley it is because of friction, which has no analogy in interplanetary space.

If simple capture is a highly improbable explanation for the moon, it would be totally impossible to believe that the dozens of other satellites in the solar system were also captured. It is far more reasonable to believe that moon formation is a natural part of planet formation. That aesthetic nicety, however, is about all the double-planet hypothesis has going for it. Physicists doubt that under normal circumstances any moon formed next to the Earth could evolve into an orbit of the present tilt. Furthermore, one might expect two bodies in the same orbit about the sun to be made of the same materials. Instead, the Earth is much denser than the moon, and I have already described other compositional differences between the two.

Currently theorists are attempting to amalgamate the better features of all three ideas, combined with new insights about the early solar system. It seems to me that an important element of any future model for the origin of the moon will be the intermixing of material throughout the early solar system. Moonlets, asteroids, comets, and other planetesimals left over from the later stages of the formation of Jupiter, Venus, and the other planets were flung about the solar system. The asteroids, grinding themselves down by repeated collisions, generated great quantities of dust that spread through the solar system. The later stages of the formation of the Earth and the formation of the moon must have been greatly influenced by all this material. Because of the differing sizes and gravities of the Earth and the moon, it would not be surprising that different percentages of high- and low-temperature condensates were incorporated in the two.

The moon may even be largely composed of one or more bodies that came whizzing past the Earth in the earliest epochs. If a pass were close enough, which is quite probable, great tidal forces would have torn such a body apart. Some of the pieces would then have gone into orbit around the Earth while the rest flew off, never to return. If such a body had already melted, formed a core, and solidified, the core might have been among the portions that continued on their way, which would account for the lack of iron-related chemicals in the moon without requiring them to be in the Earth's core. Alternatively, one can imagine collisions occurring in near-Earth space between some protolunar material and intruders from elsewhere. Because stone fractures so easily, while metal does not, there could be a net separation of the metal from the stone that ultimately resulted in the coagulation of a metal-depleted moon.

There is also new evidence of a close genetic link between the moon and the Earth. It comes from a research tool developed after the Apollo program ended. Rocks are analyzed for the ratios of the two less common isotopes of oxygen to common ^{16}O (the number is the atomic weight). Isotopes differ in the properties of their nuclei, such as their atomic weights, but not in their chemistry. So ratios of the three isotopes should remain intact, even if major chemical processes enrich or deplete a certain material in oxygen. There are just the slightest separations of the isotopes caused by the slight differences in weight. Any such change in the ratio of ^{18}O to ^{16}O should be just twice the change in ^{17}O to ^{16}O, since the difference in weight is twice as great. Yet when University of Chicago geochemist Robert Clayton measured the isotope ratios for various rocks from the Earth, the moon, and the meteorites, he found half a dozen different groupings that cannot be derived from each other by any chemical or physical process subsequent to the formation of these isotopes deep inside a star.

So there are half a dozen different places in the early solar nebula that never were thoroughly mixed. And it so happens that both the Earth and the moon are made of materials from the same part of the nebula, along with a few types of meteorites. Clayton's evidence suggests that the moon and the Earth have always been associated and that most of the material composing the moon certainly was not captured from far distant parts of the solar system. If we now can just find where those moon- and Earth-like meteorites are coming from, we could learn the size of the Earth's zone of the nebula. These particular meteorites aren't exactly moonlike in their chemistry, so they don't come from the moon itself. Perhaps their parent bodies are in the asteroid belt; the spectra of some asteroids look tantalizingly like those of the meteorites. Or maybe they come from small asteroids much closer to the Earth. In that case Mars, which is between Earth and the main asteroid belt, could have formed from its own pot of nebular material. These fascinating questions about the creation of the planets are being answered, thanks to measurements of some rocks that fall freely from the skies, of others that were returned from expensive lunar expeditions, and of still others to be returned one day from missions still on NASA's drawing boards.

Meanwhile, lunar researchers have already thought of new experimental techniques to apply to the lunar rocks that will further constrain our hypotheses for the origin of that pale disk that mankind has wondered about since first emerging on Earth. After centuries, even millennia, of theorizing about how our moon was formed, we are within a few years, or a few decades at most, of finally knowing the answer.

9

Mars and Earth: Changing Perspectives

A heavy, perhaps fatal, blow was delivered today to the possibility that there is or once was life on Mars. The Mariner 4 photographs taken of the planet at close range July 14 show a crater-pocked landscape lacking any sign that there has been water erosion there.

—*The New York Times,* JULY 30, 1965

As LATE AS the early 1960s astronomers thought Mars might be quite conducive to life. Any doubts they had sprang from their narrow conception of the tolerance of any Martian life forms rather than from the probability of any especially threatening traits in the Martian environment. (Many biologists, more familiar than astronomers with the remarkable diversity of life, have been even more optimistic about the prospects for Martian life.) Mars seemed a pleasant world. Its air was a bit thin by our standards and it got chilly at night. But there were fluffy white clouds, there were seasons like our own, and the Martian day was just 37 minutes longer than Earth's. Polar snow caps melted in the spring and dust storms occasionally blew across the planet, just as in Oklahoma. Flashes of light on Mars, witnessed by Japanese astronomers, were thought to be volcanoes, or maybe even the sun reflected from Martian lakes. Indeed there were signs of life itself. Colors of certain regions changed with the seasons, sometimes becoming bright green. And after a dusting from a passing storm, green patches sometimes reemerged, just like plants poking up through the dust to reach the sunlight. Most conclusive of all, some astronomers thought, was the spectrum of those patches that William Sinton, now of the University of Hawaii, interpreted as suggesting something like chlorophyll on Mars.

Some prescient theorists had other notions. As long ago as 1951 Ernst Öpik had predicted that "the surface of Mars should be covered with hundreds of thousands of meteor craters exceeding in size the Arizona [Meteor] Crater." And a few years later Dean McLaughlin propounded a theory to explain the changing dark patches in terms of volcanic ash being blown about by Martian winds. Öpik's sound logic was ignored and McLaughlin's theory was "disproven." We now know that much of Mars is heavily cratered and that the dark markings are caused by windblown dust, although that dust is probably not chiefly ash from the huge volcanoes that certainly are there.

A few years later Sinton realized that his spectrum was of a whiff of gas in the Earth's atmosphere rather than of plants on Mars. Still, as Mariner 4 approached Mars in summer 1965 most astronomers believed there might be lowly plants on Mars, such as the lichens that encrust desert rocks on Earth.

The men in charge of interpreting Mariner 4's 15 photographs were not planetary astronomers. One was a physicist, two were geologists (all professors at the California Institute of Technology), for the interpretation of aerial photographs of landforms is geologists' work, not the specialty of astronomers. What the pictures showed was shocking to the three, who had no previous reason for faulting the astronomers' interpretations. Ernst Öpik wasn't surprised, but almost everyone else agreed that Mars was a very different world from what had been imagined. The best pictures from Mariner 4 showed a bleak, cratered surface. Of course no photograph could have revealed life itself, unless Martians were several miles across! But the pictures might have shown river deltas, mountain chains, and other evidence of a young, active world like our own. They might even have shown canals built by a Martian civilization.

Instead, our own dead moon, whose surface was then believed —correctly—to be between 2 and 5 billion years old, was the closest likeness we had yet seen to the Mars of Mariner 4. So the Caltech researchers concluded that "the visible surface is extremely old and that neither a dense atmosphere nor oceans have been present on the planet since the cratered surface was formed." Either an ocean or an atmosphere might well have eroded any preexisting large craters, just as on Earth. But were the craters really ancient? That could be determined, roughly, by the technique of crater counting, and three separate groups of scientists immediately challenged the Caltech interpretation of crater ages.

Since they were not astronomers and had no stake in earlier views of Mars, the Caltech professors and other geologists who began studying Mars did little to sort out the significance of the changing colors, the flashes of light, and other earlier evidence of a more hospitable planet. To this day it is not really known which earlier observations were valid and which were illusions. Perhaps they hint at further surprises yet to come. The astronomers have now gone on to study other planets in the solar system, leaving Mars to the geologists.

The exploration of Mars was chiefly guided by one of the original Mariner scientists—Caltech geologist Bruce Murray, who is now the director of the Jet Propulsion Laboratory, which runs planetary space missions for NASA. He, more than anyone else, established the strategy of photographic reconnaissance of the planets. In the early 1960s, with NASA already contemplating a Mars lander project (more ambitious and called Voyager then),

Murray argued that the meager pre-Voyager reconnaissance program then planned was totally inadequate. We would be landing on Mars blind and ignorant. Suppose, for example, that Venusian crocodiles wanted to search for Earthling relatives. Their approach should *not* be to fly a Mariner 4–type spacecraft to take 15 pictures of Earth (probably showing ocean) and then drop an expensive lander onto Earth bearing crocodile goodies. Even if their lander didn't drop in the drink, it would be unlikely to land anywhere near a crocodile. A better strategy would be to send a series of reconnaissance spacecraft: the first ones would photograph Earth from distant orbit, revealing the oceans, continents, and mountains; then more sophisticated cameras would zoom in on those border regions between water and land, searching for telltale signs of swamps. They would confidently drop a costly lander only when they had first found a choice swamp.

Murray realized that the then-planned reconnaissance of Mars would have yielded only 1/10,000, or at most 1/1,000, of the information necessary to land a life detector intelligently. Under his leadership Mariners 6, 7, and 9 ultimately made up much, but not all, of the deficit in information. NASA still skipped over the highest magnification zoom-in phase of exploration. That is what caused so much consternation and delay in landing the first Viking on Mars, originally scheduled to coincide with bicentennial celebrations on July 4, 1976. Belatedly realizing that Viking could have impaled itself on never-seen boulders when it settled onto the Chryse plain, project scientists postponed the landing again and again as they desperately sought, from really inadequate data, the safest landing site.

Bruce Murray's "Martian horror story" of 1964 turned into the astounding success of Mariner 9 in 1971, which showed that the crater-scarred landscape of 1965 was not at all typical and that there might be places where Martian life thrived, after all. With hindsight we can now recognize in the pictures from Mariners 4, 6, and 7 telltale signs of unmoonlike geology that we only really noticed once Mariner 9 had photographed all of Mars. The first half-dozen pictures from Mariner 4 showed few craters on a flat and featureless terrain. We now know there are wide craterless plains in that area. But in 1965 the lack of craters was blamed on the great distance from which the pictures were taken and on the poor lighting angle. (Any amateur photographer knows one never gets good pictures aiming directly away from the sun, as was true for the first Mariner 4 frames.) Some landscapes unique to Mars were seen in a few of the pictures from Mariners 6 and 7, flown in 1969 with much improved cameras. Still, Mariners 4, 6, and 7 together photographed only a small percentage of the Martian surface at close range, so it could not have been suspected that the few unusual regions would turn out to be the headwaters of an immense system of now-dry river valleys on Mars.

With its sister spacecraft, Mariner 8, already at the bottom of the Atlantic Ocean, Mariner 9 arrived at Mars on November 14, 1971, to find the entire planet hidden beneath a pall of dust. Bruce Murray and the Mariner 9 television research team stared gloomily at each successive blank photograph from their Mars-orbiting television camera. They had been surprised and disappointed 7 weeks earlier, when Earth-based astronomers reported what was to them a familiar happening on Mars. Telescopes revealed that a small dust cloud had grown in the Noachis area in the southern hemisphere. It had rapidly expanded and spread around the planet until in November the entire globe was shrouded. Nor was this storm mere bad luck for Mariner 9; it had been predicted to occur long before Mariner 9 had been launched.

Astronomers were bitterly familiar with the disappointments of 1956, the previous time that Mars had been simultaneously close to Earth and at the closest point to the sun in its elongated orbit. Mars can be studied every 2 years from Earth, but it is much closer on special occasions every 15 or 17 years when it is also closest to the sun. Elaborate plans to scrutinize Mars during the 1956 opportunity were thwarted by a planetwide dust storm, just as observations had been impaired during every previous such occasion back to the 1800s. Observers had long supposed that the extra solar heating of the southern hemisphere, tipped sunward at such times, generated the storms that scoured pockets of yellow dust from around the dark patches that gird the southern temperate latitudes of Mars.

If the geologists had failed to heed the astronomers' warnings in planning the strategy for Mariner 9, they had nevertheless designed a flexible mission that could outlast any dust storm. Thus the great bonus afforded atmospheric specialists, who traced Martian winds from the changing dust cloud patterns, did not come at the expense of the geologists who patiently waited for the storm to subside. They then began a photographic sequence, lasting into autumn 1972, that succeeded in mapping the whole planet.

As the dust gradually settled, Mars performed an excruciatingly prolonged striptease and gradually revealed surprises that rivaled its "moonscape" exhibition of 1965. Four huge, dark mountains poked through the dust-laden atmosphere. Each was crowned by interlocking craters that resembled not meteor-impact craters but rather the volcanic crater or caldera atop the Kilauea volcano in Hawaii. The largest of these volcanoes, Olympus Mons, is the size of the state of Colorado and exceeds in size any volcano on Earth.

The second astonishing piece of scenery to be revealed was the giant chasm later christened Valles Marineris. It is the length of the United States and so vast that one of its tributary canyons almost dwarfs the Grand Canyon. Wider than the Grand Canyon is long, Valles Marineris is dark in color and had been called the Coprates canal by astronomers decades ago. Such extraordinary canyonlands imply Martian geological forces like those in the Earth. The moonlike terrains were but a relic from Mars's ancient past, like the old cratered lake districts of Canada. The young canyons and volcanoes reestablished Mars as a geologically fascinating world like the Earth.

The most portentous discovery of all came in Mariner 9 pictures of the Chryse plains near the Martian equator. They showed meandering valleys with tributaries that looked exactly like river valleys on Earth. For once, extreme caution prevailed among the geologists responsible for interpreting the pictures. They had drawn premature conclusions before and were not to do so again on the subject of Martian water, with its pregnant implications for life. But months and years of analysis of pictures taken by Mariner 9 and more recently by the Viking orbiters have not been favorable for those who have labored to explain the channels by processes other than running water. Mariner 9's telephoto lens eventually revealed that old craters are literally covered with furrows and stream channels. It is not certain, but it seems that it once rained on a now dry and dusty Martian landscape. So even before the Vikings reached the Red Planet in 1976, there was renewed hope that life nurtured during wetter climates might have adapted itself to harsher conditions and would still be found.

Questions of Martian organisms notwithstanding, the geologists who have guided our reconnaissance of that planet are pleased to have uncovered a fabulous world to study that may help us understand planetary geological processes in general and eventually enlighten us about our own planet as well.

It is easy to understand why our perceptions of the Red Planet have vacillated every few years. Until recently we had very few data from which to learn reliably about the planet's nature. Researchers were bound to be led astray as they strained to interpret every hint from the fuzzy image of the faraway planet or from the few coarse close-ups from Mariner 4. Fictional speculations, the personalities of individual scientists, and pure chance had as much to do with our earlier ideas about Mars as did hard, reliable facts. The study of Mars has been a young science and, like any young science, its practitioners' imaginations have been free to roam, unhindered by contradictory evidence. Now, with the wealth of data from Mariner 9 and Viking, Martian science must soon mature and, if history is any guide, our intellectual approaches will change. We can anticipate whither we go by examining whence we have come; and the science of the Earth provides a recent example.

Simultaneous with the flipflops in our image of Mars, geologists' conception of our own planet underwent as great a turnabout. The emergence of continental drift and sea floor spreading—the new "global plate tectonics"—during the 1960s ranks as one of the greatest revolutions in the history of science, inasmuch as it totally undercut the cherished hypotheses of generations of geologists and geophysicists. Geology was a venerable science which had matured from a stormy youth over a century ago. By the early 1900s no science was so replete with observed fact and with adopted interpretations. For another half century the contours of the hills were measured, multitudinous fossils were classified, seismograms of earthquakes recorded, and still the fundamental assumptions of geology remained unchanged.

Fifty years after Alfred Wegener proposed that continents drift, I was taught college geology in the catacombs of the Harvard University Museum. The yellowed paintings of William Morris Davis and other past giants of Harvard's renowned Geology Department hung there to remind us—teachers and students alike—of the revered traditions of a secure science. I learned how mountains grew and how they eroded. I learned that the fixed continents grew outward from central cores by "accretion." Sediments accumulated in "geosynclines" and were somehow thrust upward to form mountains during episodes called "orogenies." The concepts were laced with jargon, which I learned, but I never quite realized that my own vague confusion about what was making it all work stemmed from the fact that nobody really knew.

The new theory of continental drift and plate tectonics has brought together many diverse subjects in the study of our planet, including the growth of continents, building of mountains, deposition of mineral ores, development of sea floor ridges, seismicity, changing sea levels, and the climatology of ancient epochs. Alfred Wegener wrote in the 1920s, long before the general acceptance of his idea: "We may assume one thing as certain: The forces which displace continents are the same as those which produce great fold-mountain ranges. Continental drift, faults and compressions, earthquakes, volcanicity, marine transgression cycles [of the oceans onto land], and polar wandering are undoubtedly connected causally on a grand scale. Their common intensification in certain periods of the Earth's history shows this to be true. However, what is cause and what effect, only the future will unveil."

Global plate tectonics is the synthesis that Wegener sought, although the underlying forces within the Earth that keep the crust churning about are still only poorly understood.

Imagine a lake, frozen over with ice on a frigid winter day. Then imagine a mighty rotor beneath the ice, churning the water vigorously, breaking the ice sheet into separate sections, or plates. The jostling might thrust one plate beneath another. Water would spurt up to fill gaps between plates, only to freeze and become fresh ice. The analogy is far from exact, yet Wegener's inspiration may have come from watching ice floes during his Greenland expeditions. The analogy may be even more relevant for several of the satellites of Jupiter, which are believed to have icy crusts and liquid water mantles.

On the Earth the crustal plates are spherical shells some 70 kilometers thick (approximately 44 miles). There are half a dozen major plates and several smaller ones. Some plates are entirely of ocean floor, while others contain continents as well. A plate of ocean floor moves along like a conveyor belt, spreading from a central crack. As it converges on another plate, one dives under the other, carrying the ocean floor rocks to depths of many hundreds of kilometers into the high-pressure furnace of the Earth's mantle, where they are consumed. At such boundaries between plates are found the deepest ocean trenches, which have been buckled downward. Immediately landward of the trenches—over the descending and melting plate—volcanoes frequently form. It is also not surprising that most earthquakes, including the devastating ones in San Francisco, Alaska, and Guatemala, happen near plate boundaries where one is being thrust underneath another.

Continents are made of relatively light rock which tends to float on the denser rocks that underlie the oceans. Continents are too buoyant to sink into the mantle, so if converging plates bring two continents together, neither submerges and they smash into each other. The buckling produces great crumpled mountain chains, such as the Alps and the Himalayas. Over the eons new plate boundaries often form, sometimes in the middle of continents. South America and Africa were parts of a single continent until it split apart about 140 million years ago to form the South Atlantic Ocean. The Red Sea and the Gulf of California are very young oceans and the valleys in East Africa may mark the beginnings of another. Meanwhile, the Pacific is narrowing. Although all these motions are measured in only centimeters per year, when the displacements accumulate for millions and billions of years they amount to many complete generations of the recycling ocean floor. The oldest oceanic crust known is less than 5 percent of the Earth's age, whereas continents have grown from ancient cores whose rocks date back $3\frac{1}{2}$ billion years.

All that I have described is now accepted fact by geologists and geophysicists. Yet it was heresy only 15 years ago, despite the abundant evidence for continental drift amassed by Alfred Wegener in 1912. Although best remembered for fitting together the jigsaw-puzzle shapes of South America and Africa, Wegener did not simply make a lucky guess, ahead of his time. The final 1929 edition of his book *Die Entstehung der Kontinente und Ozeane* ("The Origin of Continents and Oceans") is a stupendous intellectual achievement. His evidence came from the disciplines of geodesy, geophysics, geology, paleontology, and paleoclimatology. He showed that ancient climates revealed by fossils made sense only if the continents had shifted with respect to the equator, the poles, and themselves. He showed that both rock units and fossil organisms matched between continents now separated by vast oceans. Only his geodetic measurement of Greenland's actual ongoing drift seems to have been based on flimsy data. His other evidence now seems to have been generally sound, as are many of his arguments as to why the then-accepted theories were wrong.

Why, then, did it take several more decades of stumbling down the wrong path for geologists to come around? It is not that

Wegener's work was obscure and ignored at the time; his book was translated into several languages and received wide discussion and debate in the 1920s. The answer lies in the sociological nature of science. When a scientific hypothesis or model seems to work and satisfy many diverse observations, it eventually comes to be accepted as a working foundation, or paradigm. Researchers who have spent years working from such a foundation cannot casually throw it away because a single contradiction appears. It is far simpler and less unsettling to doubt the new evidence. It takes a massive amount of contrary evidence to overturn dogma in a well-developed science such as geology.

It was realized even in the 1920s that Wegener made an impressive case for continental drift. But drift was antithetical to so many cherished assumptions that when a single flaw was perceived in the theory, it was sufficient to scuttle it entirely in the minds of the majority. The flaw was mainly the inadequacy of his guesses as to what forces were driving the continental motions. He argued ineffectually that doubts about the motive force should not override the direct observations that drift was actually occurring. By the late 1950s evidence for continental drift was so overwhelming that it seems geologists adhered blindly to tradition despite persuasive evidence. Finally, in the early 1960s incontrovertible proof of spreading sea floors was found in magnetic anomalies in ocean floor rocks; and a few years later the revolution occurred as everyone jumped belatedly on the bandwagon.

The revolution in Earth science is a classic example of a process defined beforehand by Thomas Kuhn in his book *The Structure of Scientific Revolutions.* Kuhn realized the inevitability and necessity that scientists would adhere to accepted dogmas until overwhelming contrary evidence was amassed. Otherwise, if scientists were open to every alternative, their research would be disorganized and misdirected. As Kuhn wrote, "The price of significant scientific advance is a commitment that runs the risk of being wrong."

Still, if geologists had been persuaded to evaluate more objectively Wegener's arguments, we might now know much more about the Earth than we do. During the 1930s, '40s, and '50s, only a handful of researchers pursued studies that had a direct bearing on what we now regard as the fundamental questions. Most scientists could not know that they were advancing geological understanding only tangentially by establishing the contradictory framework that made revolution inevitable, and indeed overdue, by the 1960s.

With this extraordinary example in mind, it is especially important that planetary scientists choose their paradigms with care. So far, with burgeoning new information about the planets, few scientists have developed commitments to any particular point of view. That is why there have been so many flipflops in our planetary conceptions. But as the new data from moon rocks and spacecraft fly-bys become digested and synthesized, there will be attempts to fashion scenarios and models that bring everything together. This process is already beginning under the banner of comparative planetology. The attempts should be lauded, but researchers must be careful not to jump on any bandwagon before we are quite sure that we're headed down a fruitful path. For if we too hastily adopt the wrong paradigm and start the psychologically inevitable habit of closing ourselves off to plausible alternatives, decades later we may find ourselves far down the wrong road. Since science is a cultural activity, the responsibility for keeping our planetary perspectives broad will be as much that of governmental funding agencies, science teachers, and the lay public as that of planetologists themselves.

Our new perspective on the Earth's geology has had inevitable repercussions for our developing views of other rocky worlds. Many of the elements of Earth's geology that are rooted in plate tectonics, including giant rift canyons and volcano chains, exist on Mars as well. Yet others are absent, such as chains of folded mountains. When planetologists first examined the variety of landforms in the Mariner photographs, they looked for Earthly counterparts. Now that there are attempts to synthesize the data and develop a global picture of Martian geology, conceptual analogs are also being sought in the geology and geophysics of our own planet.

Mars is a world divided into two parts along an equator that is tilted 35 degrees with respect to the rotational equator. The cap of the planet containing most of the dark patches is covered with many craters. The other half of the planet lies at lower elevations and is mainly a vast plain, only lightly cratered. Rising from these plains is a great continent-sized dome, topped by the largest volcanoes. The more chaotic and picturesque landscapes are generally near the border between the two hemispheres, including the great Valles Marineris canyon and many of the largest river channels. It is clear from studies of the mesas and buttes of cratered highlands perched "offshore" from the contiguous highlands that the highlands are gradually being—or have been—eaten away by erosive processes working on the cliffs that mark the highland margins.

The nature of Martian erosive processes is of great interest. The periodic dust storms undoubtedly abrade some rocks and deposit dust and sand dunes elsewhere, but we cannot be sure how efficient such windblown erosion is. On Earth only the very driest of deserts, where it rains once a century or less, have mountains carved chiefly by the winds rather than the rains. There seem to be vast regions on Mars where winds have stripped away thick layers of sediment, thereby exhuming completely buried, older landscapes.

The role of water erosion also is uncertain. Furrows, which would qualify as river valleys on Earth, cut into many older cratered terrains in a band somewhat south of and tipped 15 degrees to the true equator. This geographical distribution may reflect the zone where it was once warm enough to rain on Mars. Or furrows may have been more widespread but are only exhumed to be seen in equatorial regions where temperatures are warm enough to generate strong winds. The channels—Martian river valleys that would dwarf many of Earth's—originate mostly in relatively low-lying cratered terrain and flow into still lower-lying plains. At their headwaters the cratered regions are undercut and slumped, as though huge amounts of subterranean ice had melted and carved out the channels, causing the lands on top to cave in, with great avalanches and mud slides. Since the channels are also near the equator, a warmer climate, combined with underground volcanic heat, may be responsible for melting ground ice.

Inasmuch as 200 million years ago all the Earth's continents composed a vast single continent, called Pangaea or Gondwanaland by geologists, it is tempting to think of the Martian cratered highlands as a continent and the low-lying plains as an ocean basin. It is not impossible that waters running from the channels once filled at least some of the lowlands, but whether or not we ultimately find ocean sediments buried beneath the Martian plains, it is certain that many plains are great expanses of molten rock that has cooled, like Earth's lavas. Many plains must also be deeply blanketed by dust and sand, blown in from elsewhere. What created the hemispherical asymmetry on Mars? At least now, gravity measurements suggest that the highlands are not all simply lighter rocks floating on denser lowland rocks, as on Earth. Perhaps an eventual understanding of the Martian dichotomy will provide us a clue about the origin of our own continents.

The most interesting region on Mars is that of the great volcano-topped Tharsis dome and adjoining channel-cut lowlands to the east. The great Valles Marineris is cut into a plateau joining these two provinces. Some scientists have been tempted to view this region as one of incipient plate tectonics on Mars, since similar features exist in East Africa, where continental separation may be starting. But most researchers think it is premature to conclude that Valles Marineris represents a wrenching apart of the Martian crust and point to numerous difficulties with the analogy.

Even more debatable are possible examples of actual plate motions on Mars. For instance, on Earth the Hawaiian Islands form a long chain of volcanoes stretching across the Pacific. It has been inferred that the continual southeastward displacement of Hawaiian volcanism is due to the northwestward motion of the Pacific plate across a "hot spot" in the Earth's mantle. It is thus suggestive that several of the largest Martian volcanoes are arranged in a line. There is little additional supportive evidence for such motions on Mars, however, and it must be coincidental that the apparent displacement would just account for the inexplicable difference between the equator and the boundary between the highlands and the plains.

While exact analogies between the crusts of Earth and Mars are not justifiable, it still seems fair to say that Mars is a planet in an intermediate state of geological activity between the moon, which has long been dead, and the Earth, which is in a state of continual agitation. The moon is warm only at its center, yet the interior of Mars must be hot to have generated the vast volcanoes in relatively recent epochs. The lunar crust is thick and rigid; the Martian crust is probably much thicker than the Earth's, yet the crustal surface has been yanked apart and fractured in many places. Perhaps the Martian interior is as active as Earth's, yet it lacks a slippery layer such as that on which the Earth's plates glide.

I described in an earlier chapter how the heat generated and retained within the interiors of planets drive the geological evolution of their surfaces. It is not surprising that Mars, which is larger than the moon but smaller than Earth, is also intermediate in its geological complexity. The manner in which this planetary energy is organized in ways that affect the surfaces of planets we live on and visit is just now becoming clarified. On the Earth it is manifested in plate tectonics. For Mars, we are only now assimilating the data that will answer for us how a less vigorous planet behaves.

In the final analysis, however, it may be the superficial churnings of the atmosphere, and the water and dry ice stored in polar and underground reservoirs, that will define our ultimate image of Mars. After all, we take the mountains and lands for granted on a planet, but without air and water it would be an inhospitable place. When did it rain on Mars? Was there once an ocean? Are the water and oxygen now mainly gone from the atmosphere of Mars locked up as underground ice, or in the rusted rocks, or did they escape into interplanetary space never to return? If we have future missions to Mars, and a few million dollars a year to pay the scientists to analyze the data, we may soon know the answers to these questions just as certainly as new Martian enigmas will surely appear.

What we learn about Mars, other rocky worlds, and about our own habitat we learn through the individual and collective perspectives of scientists. We can truly appreciate our evolving knowledge of the planets only to the extent that we understand that sociological process called science that has brought us that knowledge. Although I am a practicing scientist, the process whereby men and women acquire scientific wisdom is as marvelously mysterious to me as such other human endeavors as politics. Thus I have been able to do little more than provide occasional anecdotes amid my descriptions of planetary science from the perspective of a practitioner.

Yet it seems to me an intellectual accomplishment of the most awe-inspiring sort that in the few centuries since Galileo invented the telescope we learned so much about the planets merely by studying their faint shimmering light reaching us across the vastness of space. And it is a technological marvel of the Space Age that we now study pieces of the moon in our laboratories or direct our remote-controlled robots sitting on ground millions of miles away to hammer at a Martian rock.

Index